RELATIVISTIC QUANTUM FIELDS

Charles Nash

Department of Mathematical Physics
National University of Ireland
Maynooth, Ireland

DOVER PUBLICATIONS, INC.
Mineola, New York

Copyright

Copyright © 1978 by Charles Nash
All rights reserved.

Bibliographical Note

This Dover edition, first published in 2011, is an unabridged republication of the work originally published in 1978 by Academic Press, Inc., New York.

International Standard Book Number
ISBN-13: 978-0-486-47752-7
ISBN-10: 0-486-47752-5

Manufactured in the United States by Courier Corporation
47752501
www.doverpublications.com

Preface

In the last ten years or so there have been a number of attempts to make progress in various problems in quantum field theory. This has resulted in the creation of several new techniques which are now important for a research worker doing calculations in field theory. Among the techniques that we have in mind are: the short distance expansion of Wilson, the dimensional regularization method, and the renormalization group methods.

This book is an attempt to present these techniques in a form comprehensible to post-graduate students who are in the latter half of their first year of research. The two topics which receive most attention are the dimensional method and the renormalization group method. However there are also chapters on functional integration and differentiation. Functional techniques are also used to discuss the gauge properties of QED. The infrared problem of QED is also dealt with in this book. For the most part, two field theories are discussed: $\lambda\phi^4/4!$ theory and QED. There is, however, a section on Yang-Mills theories in the last chapte as such theories are regarded as being important both experimentally and theoretically nowadays.

As preliminary reading for this book the early chapters of the following books are suggested:

J. D. Bjorken and S. D. Drell, "Relativistic Quantum Mechanics" McGraw-Hill, 1964.

S. Schweber, "An Introduction to Relativistic Quantum Field Theory", Harper and Row, New York, 1961.

R. J. Eden, P. V. Landshoff, D. I. Olive and J. C. Polkinghorne, "The Analytic S-Matrix", Cambridge University Press, 1966.

Although some of the sections of this book are fairly self-contained the reader will find it more convenient to read topics in approximately the order in which they appear in the text.

June 1978 CHARLES NASH

Contents

Preface v

1. Renormalization, Functional Differentiation and Integration, and the Schwinger–Dyson Equations
Renormalization of $\lambda\phi^4/4!$ Theory 1
Renormalization of QED 9
Functional Methods and Path Integrals 15
Functionals and Fermions 19
Functional Taylor Series 23
Functional and Path Integrals 24
Functional Integrals for Fermions 45
Generating Functions 50
The Schwinger–Dyson Equations 55
References 60

2 Dimensional Regularization and $\lambda\phi^4/4!$ Theory
Introduction 61
Dimensional Regularization and $\lambda\phi^4/4!$ Theory . . 63
Renormalization in General 86
A Counter Example 89
References 90

3. Dimensional Regularization of Quantum Electrodynamics
Quantum Electrodynamics: Lagrangian and Feynman Rules 91
Renormalization of QED 92

The Ward Identity	95
Dimensional Method for QED	100
Computation in D Dimensions in QED, Ward Identity Verification and Evaluation of Renormalization Constants	104
Shortcoming of the Dimensional Method: The ε Symbol and the Triangle Anomaly	118
Criterion for Renormalizability in General	121
References	124

4. The Gauge and Infrared Properties of Quantum Electrodynamics

Gauge Fixing Terms	125
The Ward Identities	135
An Instructive Example	139
Ward Identity and Unitarity	141
Infrared Divergences of QED	148
Coherent States	150
Cancellation of Infrared Divergences	153
References	161

5. Asymptotic Behaviour and Renormalization Group Methods

Asymptotic Behaviour of Scattering Amplitudes	163
Renormalization Group Methods	164
The Gell-Mann–Low Equations	165
The Callan–Symanzik Equations	166
The Wilson Expansion at Short Distances	188
Infrared and Ultraviolet Stability, Fixed Points	190
Asymptotic Freedom	194
The Callan–Symanzik Equations for QED	203
The Connection between Zero Mass Theories and Γ_{as}	203
The 't Hooft—Weinberg Equations	204
Non-leading Asymptotic Behaviour from the Callan–Symanzik Equations	209
Minkowskian versus Euclidean Momenta	211
References	212

Appendix	214
Bibliography	219
Index	221

ONE

Renormalization, Functional Differentiation and Integration, and the Schwinger–Dyson equations

Renormalization of $\lambda\phi^4/4!$ Theory

The quantum field theories that we shall be concerned with in this book can all be described in terms of a Lagrangian L. The two theories that we shall discuss in detail and use for providing examples of calculational techniques are the theory of a massive, (spinless), self-interacting scalar particle†: (we shall call this $\lambda\phi^4/4!$ theory after a term appearing in its Lagrangian and shall often use this symbol to denote this particular theory), and the theory of electrons and positrons interacting with photons: quantum electrodynamics (usually abbreviated as QED). However, before immersing ourselves in the study of the interactions of the particles represented by these Lagrangians, let us first make contact with the free particles which result in the idealized world where no particles interact. Beginning with the massive scalar particle we have the free Lagrangian L_F given by

$$L_F = \tfrac{1}{2}(\partial_\mu\phi)(\partial^\mu\phi) - m^2\phi^2/2 \qquad (1.10)$$

† This theory is chosen because of its simplicity: absence of spin and internal quantum numbers and the convenience of having only one particle type which interacts with itself. It is not intended to be taken seriously in an experimental sense as is quantum electrodynamics.

Given a Lagrangian like this a mathematical physicist naturally writes down the variational equations which render the action

$$S = \int d^4x L \tag{1.11}$$

an extremum. An action principle for quantum mechanics is a familiar and very useful concept developed especially by Schwinger.[1] The variational equation is

$$\frac{\partial}{\partial x_\mu}\left(\frac{\delta L}{\delta \partial_\mu \phi(x)}\right) - \frac{\delta L}{\delta \phi(x)} = 0. \tag{1.12}$$

Applying this to the Lagrangian L_F gives the well known Klein Gordon equation

$$(\Box + m^2)\phi(x) = 0. \tag{1.13}$$

This equation may be solved directly by Fourier transformation. The next step is to include an interaction term in the Lagrangian. To this end we add to the free Lagrangian L_F the interaction term $L_I = -\lambda \phi^4/4!$. The final form of the Lagrangian is thus

$$L = L_F + L_I = \tfrac{1}{2}(\partial_\mu \phi)^2 - \frac{m^2 \phi^2}{2} - \frac{\lambda \phi^4}{4!}. \tag{1.14}$$

The factor 4! is included for combinatorial reasons only. This will be illustrated by the calculations of Chapter 2. Proceeding in analogy with the free case we obtain the variational equation

$$(\Box + m^2)\phi = -\frac{\lambda \phi^3}{3!}. \tag{1.15}$$

We have no general method for solving this equation. Its non-linear character is a great headache. The best that we can do as an attempt to progress further is to linearize the problem. Fortunately this is something that we know how to do and we already have a technique at our disposal. The technique is, of course, the perturbation expansion of Feynman diagrams for the Lagrangian 1.14. The reader is expected to be familiar with the Feynman diagrams for QED and perhaps those for pseudoscalar meson theory, but not those for $\lambda \phi^4/4!$ theory. Due to the relative simplicity of the Lagrangian of 1.14 the Feynman diagrams for $\lambda \phi^4/4!$ theory are straightforward to master. We shall now describe them. There is one kind of vertex and only one kind of internal or external line, examples of these are shown in Fig. 1.

(a) vertex (b) graph with internal and external lines

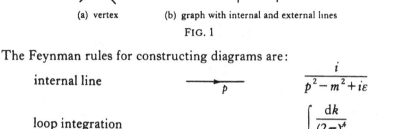

FIG. 1

The Feynman rules for constructing diagrams are:

internal line		$\dfrac{i}{p^2-m^2+i\varepsilon}$
loop integration		$\displaystyle\int \frac{dk}{(2\pi)^4}$
vertex	$p+q+r+s=0$	$-i\lambda$
symmetry factor	S	(1.16)

In this chapter we shall ignore the symmetry factor S. It will be discussed in detail in the next chapter. We now wish to give an elementary account of how renormalization occurs in $\lambda\phi^4/4!$ theory to remind the reader of some of the features of renormalization. We expect the reader to be familiar with only the gross features of renormalization; no detailed knowledge is taken for granted.

Our task is now to examine a simple graph to see how renormalization occurs. The simplest graph that we can take is one with two external lines, i.e. the propagator graph. To lowest order it is just a single unadorned line

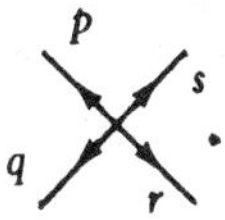

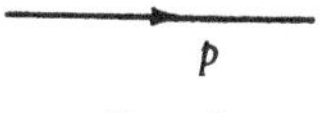

FIG. 2

representing the expression $+i/(p^2-m^2+i\varepsilon)$. To next order in the coupling constant λ it is given by the sum of two graphs†

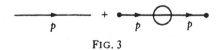

FIG. 3

† The black dots in Fig. 3 draw attention to the fact that propagators are attached to the external lines, cf. eq. 1.17.

The expression we now have is

$$\frac{i}{p^2-m^2+i\varepsilon} + \left\{\frac{i}{(p^2-m^2+i\varepsilon)}\right\}^2$$

$$\cdot \int \frac{dk_1\, dk_2}{(2\pi)^8} \frac{(-i\lambda^2)}{(k_1^2-m^2)(k_2^2-m^2)\{(p+k_1-k_2)^2-m^2\}}$$

$$= \frac{i}{(p^2-m^2+i\varepsilon)}\left\{1+\frac{i\Pi(p)}{(p^2-m^2+i\varepsilon)}\right\} \quad (1.17)$$

where $\Pi(p)$ stands for the integral over k_1 and k_2. Now if we want to identify the mass of the particle, things are not so straightforward. To lowest order in λ, the mass was simply the position of the pole in p^2 of the propagator. But the second order expression 1.17 may have a pole and a double pole at $p^2 = m^2$ if $\Pi(p)$ is regular there. To see what really does describe the true state of affairs we must take the calculation a little further. What we do is to make a model for the propagator in which the graph represented by Π is iterated an unlimited number of times. This is represented graphically in Fig. 4

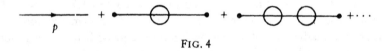

FIG. 4

The series of graphs in Fig. 4 is our model for the propagator. In mathematical quantities it is simply the series

$$\frac{i}{(p^2-m^2)}\left\{1+\frac{i\Pi}{(p^2-m^2)}+\left(\frac{i\Pi}{(p^2-m^2)}\right)^2+\cdots\right\}$$

$$= \frac{i}{(p^2-m^2)} \frac{1}{1-[i\Pi/(p^2-m^2)]}$$

$$= \frac{i}{p^2-m^2-i\Pi}. \quad (1.18)$$

Now we can see that because $\Pi(p)$ is not zero the position of the pole is no longer m^2. The new pole will be at the value of p^2 for which

$$p^2 - m^2 - i\Pi(p) = 0. \quad (1.19)$$

This value of m is called the renormalized mass m_R. Thus our simple, model shows that the presence of interaction shifts the mass from the value m to the value m_R. Note that this renormalization takes place simply because of the presence of the interaction and has nothing to do, *a priori*, with infinite quantities. It is well known that infinite quantities do, however, arise in renormalizing most quantum field theories. It will be the task of subsequent chapters to deal with some of the problems and properties of these infinite quantities. The next thing to inquire of this pole at m_R^2 is what is the value of its residue? To answer this question we assume that the function $\Pi(p)$ is regular in the neighbourhood of $p^2 = m_R^2$ and expand it about that point. We shall only need the first few terms of the series, and so we write

$$i\Pi(p^2) = \Pi_1 + (p^2 - m_R^2)\Pi_2 + \frac{(p^2 - m_R^2)^2}{2}\Pi_3 + \cdots \qquad (1.20)$$

Evidently when $p^2 = m_R^2$, $i\Pi(p^2) = \Pi_1$; thus we may write

$$m^2 + \Pi_1 = m_R^2. \qquad (1.21)$$

The form of 1.18 is now

$$\frac{i}{p^2 - m_R^2 - (p^2 - m_R^2)\Pi_2 - [(p^2 - m_R^2)^2/2]\Pi_3 + \cdots}. \qquad (1.22)$$

It is clear then that as the value of $p^2 \to m_R^2$ the residue of the pole is $+i/(1 - \Pi_2)$ where Π_2 may be written

$$\begin{aligned}\Pi_2 &= i\frac{\partial \Pi(p)}{\partial p^2} \\ p^2 &= m_R^2.\end{aligned} \qquad (1.23)$$

This constant $(1 - \Pi_2)^{-1}$ is called the wave function renormalization constant and is conventionally written as Z_3. In terms of Z_3 and m_R^2, i.e. in terms of the renormalized quantities, the propagator has the structure (we now delete the factor i to conform with standard practice)

$$\frac{Z_3}{p^2 - m_R^2 - Z_3(p^2 - m_R^2)^2(\Pi_3/2) + \cdots} = D(p, m_R^2). \qquad (1.24)$$

The value m_R^2 of the mass is taken to be the true physical value of the mass and the quantity m^2 is usually written m_0^2 and called the bare mass. Also the residue at the physical value m_R^2 is taken to be the definition of the wave function renormalization constant Z_3. It is also conventional to define what is called the renormalized propagator D_R by

$$D_R = Z_3^{-1} D. \qquad (1.25)$$

Thus

$$D_R = \frac{1}{p^2 - m_R^2 - Z_3[(p^2 - m_R^2)^2/2]\Pi_3 + \cdots} \qquad (1.26)$$

and D_R has residue unity. What this amounts to is that we renormalize the propagator to have the same scale as the bare propagator. This procedure can be translated into the language of fields. When doing second quantization we learned that the propagator was found by taking the vacuum expectation value of the time ordered product of two fields. In momentum space this amounts to writing

$$D(p) = -i \int dx\, e^{-ipx} \langle 0 | T(\phi(x)\phi(0)) | 0 \rangle$$
$$= \frac{Z_3}{p^2 - m_R^2 - Z_3[(p^2 - m_R^2)^2/2]\Pi_3 + \cdots}. \qquad (1.27)$$

Let us introduce the renormalized field $\phi_R(x)$ defined by

$$\phi_R = Z_3^{-1/2} \phi$$

(ϕ is called the bare field and often written ϕ_0). Then the renormalized propagator $D_R(p)$ is simply obtained by replacing the bare quantities by their renormalized counterparts. In other words,

$$D_R(p) = -i \int dx\, e^{-ipx} \langle 0 | T(\phi_R(x)\phi_R(0)) | 0 \rangle. \qquad (1.28)$$

It is advantageous to work with the renormalized quantities ϕ_R and m_R rather than the bare ones ϕ_0 and m_0. This really only becomes convincing when one has got down to the bread and butter task of actually doing calculations such as those described in chapter two.

There is a further renormalization which we must discuss in $\lambda \phi^4/4!$ and that is vertex renormalization which leads one to a renormalized coupling λ_R. This is quite simple to approach. The scattering amplitude that is appropriate to consider is the "two to two amplitude", i.e. those Feynman graphs with exactly four external legs.

The first four graphs are shown in Fig. 5. The lowest order graph is simply the single vertex representing the expression $-i\lambda$. The next three graphs are of order λ^2 and we lump these together and call their contribution $-i\lambda^2 \Gamma_1(p, q, r, s)$. To this order then the content of Fig 5 may be expressed in the equation

$$\Gamma(p, q, r, s) = \lambda + \lambda^2 \Gamma_1(p, q, r, s) + \cdots . \tag{1.29}$$

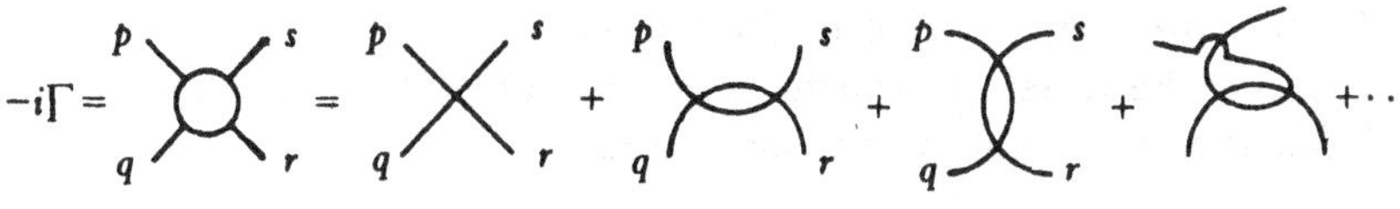

FIG. 5

We now raise the question of what we mean by the coupling constant? Obviously we would like a definition which embodies something physical and perhaps measurable. We intuitively feel that the function $\Gamma(p, q, r, s)$ measures the strength with which four particles interact, and surely this must have something to do with the concept of a coupling constant? However the parameter λ which we have up to now thought of as the coupling constant does not look very satisfactory for this role since it occurs in Γ as a power series. The way out of this puzzle is to *define* λ', another coupling constant, to be the number obtained when Γ is evaluated at some agreed values of p, q, r, s. We adhere to convention in choosing these values and define λ' by

$$\lambda' = \Gamma(p_i), \quad p_i = \hat{p}_i \qquad \hat{p}_i \cdot \hat{p}_j = (4\delta_{ij} - 1)\frac{m^2}{3} \tag{1.30}$$

where $p_1 \cdots p_4$ are $p, q, r,$ and s respectively. This is sometimes called the symmetry point because of the symmetry with which it treats the momentum variables $p_1 \cdots p_4$. So to lowest order λ' and λ are equal while to next order we have

$$\begin{aligned}\lambda' &= \lambda + \lambda^2 \Gamma_1(\hat{p}_i) \\ &= \{1 + \lambda \Gamma_1(\hat{p}_i)\}\lambda \end{aligned} \tag{1.31}$$

We now define the constant Z_1, the vertex renormalization constant, so that eq. 1.31 reads

$$\lambda' = Z_1^{-1}\lambda$$

where

$$Z_1 = \frac{1}{1 + \lambda \Gamma_1(\not{p}_1) + \cdots}. \tag{1.32}$$

To get the actual definition of the renormalized coupling constant we return to the Langrangian L, and recall that the interaction term is $-\lambda \phi^4/4!$; in terms of λ', this becomes $-Z(\lambda' \phi^4/4!)$. In terms of λ' and ϕ_R the expression is $-Z_1 Z_3^2 (\lambda' \phi_R^4/4!)$. We wish to absorb the Z_3^2 factor and leave the vertex renormalization constant by itself. To this end we define the renormalized coupling constant λ_R by

$$\lambda_R = Z_1^{-1} Z_3^2 \lambda \tag{1.33}$$

with λ the bare coupling constant. The interaction term $-\lambda \phi^4/4!$ written in terms of λ_R and ϕ_R becomes simply

$$-Z_1 \lambda_R \frac{\phi_R^4}{4!}. \tag{1.34}$$

Let us make a list of our renormalized quantities, so far we have

$$\begin{aligned} \phi_R &= Z_3^{-1/2} \phi_0 \\ \lambda_R &= Z_1^{-1} Z_3^2 \lambda_0 \\ m_R^2 &= m_0^2 + \Pi_1. \end{aligned} \tag{1.35}$$

In fact, on a point of convention, the symbol Π_1 is never used. Instead the symbol δm^2 which is called the 'mass shift' is used. Also the reader will note that we have followed conventional practice and inserted the suffix 0 on the bare quantities. To complete this discussion we display the Lagrangian L of eq. 1.14, (which we now call the bare Lagrangian and write it as L_0), written using the renormalized quantities λ_R, ϕ_R, m_R and δm^2. The result is the expression

$$L_0 = \frac{Z_3}{2} \{ (\partial_\mu \phi_R)(\partial^\mu \phi_R) - m_R^2 \phi_R^2 \} - Z_1 \lambda_R \frac{\phi_R^4}{4!} + Z_3 \frac{\delta m^2}{2} \phi_R^2. \tag{1.36}$$

The careful reader will notice that we have done some juggling of terms so as to display by itself the last term $Z_3(\delta m^2 \phi_R/2)$. This will be convenient for the calculations that we shall perform in the next chapter. The foregoing discussion has not been rigorous; this should not be allowed to be a worry. The intention is to present the main ideas and motivation and not to embark upon a rigorous treatment which would become too long and difficult to follow.

Renormalization of QED

It is now a fairly straightforward matter to repeat this analysis for quantum electrodynamics. The Lagrangian is

$$L = L_F + L_I$$

with

$$L_F = -\tfrac{1}{4}F_{\mu\nu}F^{\mu\nu} - \tfrac{1}{2}(\partial_\lambda A^\lambda)^2 + \bar\psi(i\slashed\partial - m)\psi \tag{1.37}$$

and

$$L_I = -e\bar\psi\gamma_\mu A^\mu \psi, \qquad \slashed\partial = \gamma^\mu \partial_\mu. \tag{1.38}$$

There are now two kinds of particle, the photon A_μ and the electron $\psi(x)$. Each field has its own variational equation. That for the photon is, in its free form

$$-\partial_\mu(\partial_\lambda A^\lambda) + \partial^\lambda F_{\lambda\mu} = 0 \tag{1.39}$$

and

$$-\partial_\mu(\partial_\lambda A^\lambda) + \partial^\lambda F_{\lambda\mu} = e\bar\psi\gamma_\mu \psi \tag{1.40}$$

in its interacting form. The corresponding pair of equations for the electron is[†]

$$(i\slashed\partial - m)\psi = 0 \tag{1.41}$$

and

$$(i\slashed\partial - m)\psi = e\gamma_\mu A^\mu \psi. \tag{1.42}$$

The first difference between QED and $\lambda\phi^4/4!$ theory that becomes apparent is that we must, at least formally, solve a pair of coupled equations in the functions $\psi(x)$ and $A_\mu(x)$. The Feynman rules that

---

[†] Actually there is an asymmetry in the treatment of ψ and $\bar\psi$ in 1.37. We can remedy this by adding the term $\bar\psi(i\slashed\partial - m)\psi$ in which the derivative acts to the left. The difference in the two Lagrangians is only a four divergence and so should not have any physical consequence. The pair of equations 1.41, 1.42 were obtained by considering variations with respect to the field $\bar\psi(x)$.

we use to construct diagrams are

internal electron line	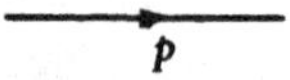 ——→—— p	$\dfrac{i}{\not{p}-m+i\varepsilon}$
internal photon line	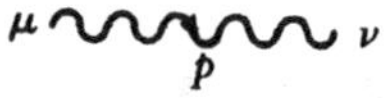 μ ⁓⁓⁓ ν p	$\dfrac{-ig_{\mu\nu}}{p^2+i\varepsilon}$
vertex	 (diagram) μ	$-ie\gamma_\mu$
closed Fermion loop		$-$trace
loop momentum k		$\displaystyle\int\dfrac{dk}{(2\pi)^4}$
external electron line		$\bar{u}(p,s)$ outgoing $u(p,s)$ incoming
external photon line		ε_μ (1.43)

For the basic definitions and properties of Dirac matrices we refer the reader to the book by Bjorken and Drell already cited in the preface. We shall prove the more detailed properties of Dirac matrices that we use in this book. In the appendix there is a summary of the basic properties.

We now wish to discuss the renormalization of the photon field. The mass renormalization for the photon field should not in this case change the mass of the photon since we want this to remain zero. This is guaranteed by gauge invariance and will be pursued in detail in Chapter 4 so we do not pass further comment on it here. The photon will, however, receive a wave function renormalization. Let us see how this comes about. Again we take a simple model for the photon propagator $D_{\mu\nu}$, this is shown in Fig. 6.

$D_{\mu\nu} =$ ⁓⁓⁓ $+$ ⁓O⁓ $+$ ⁓O⁓O⁓

FIG. 6

We denote the simple Fermion 'bubble' graph which is being iterated by $D(p^2)$. Summing the geometrical series we arrive at the expression

$$D_{\mu\nu} = -\frac{ig_{\mu\nu}}{p^2+D(p^2)}. \quad (1.44)$$

Bearing in mind our remark about the renormalized mass of the photon we expand $D(p^2)$ about the point $p^2 = 0$

$$D(p^2) = p^2 d_1 + p^4 d_2 + \cdots . \tag{1.45}$$

Taking the limit of $D_{\mu\nu}$ as $p^2 \to 0$ we find immediately that

$$D_{\mu\nu} \underset{p^2 \to 0}{=} \frac{-ig_{\mu\nu}}{p^2(1+d_1)} \tag{1.46}$$

thus

$$Z_3 = (1 + d_1)^{-1}. \tag{1.47}$$

Turning now to the electron propagator $S_F(p)$ construct the graphical series shown in Fig. 7.

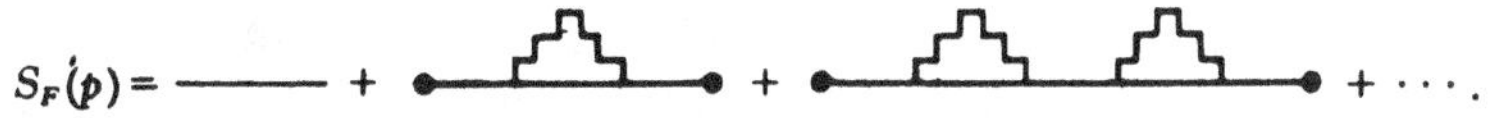

$$S_F(p) = \text{———} + \text{—●—⌐⌂¬—●—} + \text{—●—⌐⌂¬—●—⌐⌂¬—●—} + \cdots .$$

FIG. 7

We write the summed series as

$$S_F(p) = \frac{i}{\not{p} - m - \Sigma(p)} \tag{1.48}$$

with $\Sigma(p)$ defined to be the integral

$$-ie^2 \int \frac{dk}{(2\pi)^4} \frac{\text{Tr}\{\gamma_\lambda(\not{k}+\not{p}+m)\gamma^\lambda\}}{\{(k+p)^2 - m^2\}k^2}. \tag{1.49}$$

As is familiar from the preceding examples we realise that the presence of the quantity Σ in 1.48 means that the mass has been renormalized. But Σ is also a 4×4 matrix in Dirac space so we must take account of this in defining the renormalization constant and the renormalized mass. This is not difficult. Let m_R be the renormalized mass, then we must expand $\Sigma(p)$ about the point

$$p^2 = m_R^2 \tag{1.50}$$

or

$$\not{p} = m_R. \tag{1.51}$$

We shall first deal with the matrix structure of Σ. In Chapter 3 we prove that any 4×4 matrix may be expanded as a linear combination of the 16 Dirac matrices

$$\{I, \gamma_5, \gamma_\mu, \gamma_5\gamma_\mu; \sigma_{\mu\nu}, \mu \neq \nu\}, \qquad \mu = 0, 1, 2, 3$$

where

$$\gamma_5 = i\gamma^0\gamma^1\gamma^2\gamma^3, \qquad \sigma_{\mu\nu} = \tfrac{1}{2}i\{\gamma_\mu\gamma_\nu - \gamma_\nu\gamma_\mu\}. \qquad (1.52)$$

The matrix Σ may be expressed as shown below:

$$\Sigma(p) = A_1(p^2)I + A_2(p^2)\gamma_5 + A_3(p^2)\slashed{p} + A_4(p^2)\gamma_5\slashed{p} + A_5(p^2)p_\mu p_\nu \sigma_{\mu\nu}, \qquad (1.53)$$

The expansion of 1.53 contains five terms; fortunately some of these vanish. The ones which are non-zero are $A_1(p^2)$ and $A_3(p^2)$. The function $A_5(p^2)$ need not be zero but its coefficient $p_\mu p_\nu \sigma_{\mu\nu}$ is zero since $\sigma_{\mu\nu} = \tfrac{1}{2}i[\gamma_\mu, \gamma_\nu]$ is antisymmetric in μ and ν. The remaining two functions $A_2(p^2)$ and $A_4(p^2)$ are zero because their coefficients γ_5 and $\gamma_5\gamma_\mu$ have the opposite behaviour under parity to the matrices I and γ_μ. This means that $A_2(p^2)$ and $A_4(p^2)$ must be zero because QED is invariant under parity, and $\Sigma(p)$ has a fixed transformation law under parity. The parity properties of $\Sigma(p)$ must be the same as that of the 'bare term' $\slashed{p} - m$; this is why A_2 and A_4 vanish rather than A_1 and A_3. The result of all this enables us to write

$$\Sigma(p) = A_1(p^2) + A_3(p^2)\slashed{p}. \qquad (1.54)$$

Now we expand about $\slashed{p} = m_R$ to obtain the expansion

$$\Sigma(p) = A + B(\slashed{p} - m_R) + \cdots \qquad (1.55)$$

with

$$A = \Sigma(p), \qquad \slashed{p} = m_R$$

and

$$B = \frac{\partial \Sigma(p)}{\partial \slashed{p}}, \qquad \slashed{p} = m_R. \qquad (1.56)$$

Equations 1.55,56 are, of course, consistent with the form given in 1.54. Now the renormalized mass is found to be

$$m_R = m + A = m + \delta m, \qquad (1.57)$$

while in the limit $\slashed{p} \to m_R$

$$S_F(p) \to \frac{i}{(\slashed{p} - m_R)[1 + B]}. \qquad (1.58)$$

So Z_2 is defined by

$$Z_2 = [1+B]^{-1} \tag{1.59}$$

(The suffix 2 is conventional for electrons.) Having defined the renormalized parameters appropriate for the internal lines it remains for us to define a renormalization constant for the vertex. The symbol used for the vertex is $\Gamma_\mu(p', p)$ but with a factor of $-ie$ and with the external spinor wave functions removed. For example, the bare vertex of Fig. 8

FIG. 8

stands for the expression

$$-ie\bar{u}(p')\Gamma_\mu(p', p)u(p) \tag{1.60}$$

so that in this case

$$\Gamma_\mu = \gamma_\mu. \tag{1.61}$$

The vertex $\Gamma_\mu(p', p)$ is renormalized by the higher order graphs, the first one of which is displayed in Fig. 9.

FIG. 9

The mathematical expression for this diagram we shall write as $-ie\bar{u}(p')\Gamma_\mu^1(p', p)u(p)$. In terms of $\Gamma_\mu^1(p', p)$ and γ_μ the vertex is

$$\Gamma_\mu(p', p) = \gamma_\mu + e\Gamma_\mu^1(p', p). \tag{1.62}$$

Now just as we had to choose a particular value of the external momenta in $\lambda\phi^4/4!$ theory to define Z_1, so we must do likewise here. The values of momenta which we choose are

$$\begin{aligned} p &= p' \\ p^2 &= (p')^2 = m_R^2. \end{aligned} \tag{1.63}$$

This question of choice of normalization point has certain subtleties associated with it that will be fully dealt with in Chapter 3. For the moment let it suffice to say that at the normalization point specified in 1.63 $\Gamma_\mu^1(p', p)$, (and so also $\Gamma_\mu(p', p)$), becomes proportional to the Dirac matrix γ_μ

$$\Gamma_\mu^1(p', p) = e^2 C \gamma_\mu \qquad p' = p \qquad p^2 = m_R^2. \tag{1.64}$$

so that

$$\Gamma_\mu(p, p) = (1 + e^2 C) \gamma_\mu \qquad p^2 = m_R^2. \tag{1.65}$$

Equation 1.65 provides us with a definition of Z_1, the vertex renormalization constant, namely

$$Z_1 = (1 + e^2 C)^{-1}. \tag{1.66}$$

Equation 1.66 makes our list of renormalized quantities for QED complete. It remains for them to be displayed in Lagrangian form in the same way as we did for $\lambda \phi^4/4!$ theory. The variables which we want to use to write out the Lagrangian are

$$\begin{aligned} A_R^\mu(x) &= Z_3^{-1/2} A_0^\mu(x) \\ \psi_R(x) &= Z_2^{-1/2} \psi_0(x) \\ e' &= Z_1^{-1} e_0 \\ m_R &= m + \delta m \end{aligned} \tag{1.67}$$

where we have, as is usual, used the suffix 0 to denote the bare quantities. The Lagrangian of 1.37, 38 written in terms of the renormalized quantities is

$$L = \frac{-Z_3}{4} (\partial_\mu A_\nu^R - \partial_\nu A_\mu^R)^2 + \frac{Z_3}{2} (\partial_\lambda A_R^\lambda)^2 + Z_2 \bar{\psi}_R (i \not{\partial} - m_R) \psi_R$$
$$- Z_1 Z_2 Z_3^{1/2} e' \bar{\psi}_R \gamma_\mu A_R^\mu \psi_R + Z_2 \delta m \bar{\psi}_R \psi_R. \tag{1.68}$$

The last renormalized quantity, the renormalized coupling constant, tidies up the appearance of the interaction term. It is defined by

$$\begin{aligned} e_R &= Z_2 Z_1^{-1} Z_3^{1/2} e_0 \\ &= Z_2 Z_3^{1/2} e'. \end{aligned} \tag{1.69}$$

The interaction term is then written simply as

$$- Z_1 e_R \bar{\psi}_R \gamma_\mu A_R^\mu \psi_R. \tag{1.70}$$

This concludes, (until Chapter 3), our discussion of the renormalization of QED.

Functional Methods and Path Integrals

For the rest of this chapter we wish to introduce the reader to two important techniques: Functional Methods and Path Integrals. Neither of these techniques is particularly new but in recent times they have been receiving quite a lot of attention because of the interest in the so-called gauge theories of weak and electromagnetic interactions.[2] We shall deal with these methods together. One of the exciting achievements of these methods is that they allow one to derive exact equations for the Green's functions of the theory. These are called the Schwinger–Dyson equations. We shall derive the Schwinger–Dyson equations for $\lambda\phi^4/4!$ theory and QED in this chapter. The mathematical innovation which provides the foundation for these techniques is the idea of differentiation with respect to functions rather than with respect to real or complex numbers; and integration with respect to a measure defined on a space of functions rather than with respect to the Riemann or Lebesgue measure used for real and complex numbers. The former might loosely be called functional differentiation; and the latter functional integration. We begin naturally enough with functional differentiation. It transpires that we have to leave the ordinary space of functions $f(x)$ and include generalized functions, (in particular the delta functions $\delta(x-x')$), to make the definition of a functional derivative. Let $E(f(x))$ be a functional of $f(x)$ for example we might have

$$E(f) = \int dx f(x)$$

or

$$E(f(x)) = f(x), \quad x \text{ fixed.} \tag{1.71}$$

The definition of the functional derivative is as follows: let ε be a real positive number, then the functional derivative of $E(f)$ with respect to $f(y)$ is written $\delta E/\delta f(y)$ where

$$\frac{\delta E(f(x))}{\delta f(y)} = \lim_{\varepsilon \to 0} \frac{E(f(x)+\varepsilon\delta(x-y))-E(f(x))}{\varepsilon}. \tag{1.72}$$

To make things clearer we compute the functional derivatives of the two examples of $E(f(x))$ given in 1.71. When $E(f(x)) = f(x)$ then

$$\frac{\delta E(f(x))}{\delta f(y)} = \lim_{\varepsilon \to 0} \frac{f(x) + \varepsilon \delta(x-y) - f(x)}{\varepsilon} = \delta(x-y) \quad (1.73)$$

1.73 is a rather convenient formula since it enables us to compute our second functional derivative. When $E(f) = \int dx\, f(x)$

$$\frac{\delta E(f)}{\delta f(y)} = \lim_{\varepsilon \to 0} \frac{\int dx\, (f(x) + \varepsilon \delta(x-y)) - \int dx\, f(x)}{\varepsilon}$$

$$= \int dx\, \delta(x-y)$$

$$= 1. \quad (1.74)$$

However more simply we can make use of 1.73 to write

$$\frac{\delta E(f(x))}{\delta f(y)} = \int dx\, \frac{\delta f(x)}{\delta f(y)}$$

$$= \int dx\, \delta(x-y)$$

$$= 1. \quad (1.75)$$

Another example is provided by the kernel term from a typical linear integral equation, i.e.†

$$E_x(f) = \int dx'\, K(x, x') f(x'). \quad (1.76)$$

It is then straightforward to compute that

$$\frac{\delta E_x(f)}{\delta f(y)} = K(x, y). \quad (1.77)$$

There are some reasonable similarities to ordinary differentiation, and the well known chain rule for functions of a function has a functional counterpart. Let $E = E(F)$ and $F = F(f(x))$, then the chain rule we

† In 1.76 x is a parameter and not an independent variable. This is why the LHS is written $E_x(f)$ rather than the less correct $E(f(x))$.

seek says that

$$\frac{\delta E}{\delta f(y)} = \int dx' \frac{\delta E}{\delta F(x')} \frac{\delta F(x')}{\delta f(y)}. \qquad (1.78)$$

An example of the use of the rule 1.78 is provided by choosing $E(F) = F$ and $F(x) \equiv F(f) = \int dx_1 \, dx_2 \, f(x_1)f(x_2)$. We find that

$$\frac{\delta E(x)}{\delta F(x)} = \delta(x - x')$$

and

$$\frac{\delta F(f)}{\delta f(y)} = 2 \int dz \, f(z), \qquad (1.79)$$

so the chain rule tells us that

$$\frac{\delta E(x)}{\delta f(y)} = 2 \int dx' \, \delta(x - x') \int dz \, f(z)$$
$$= 2 \int dz \, f(z). \qquad (1.80)$$

Alternatively, we can simply compute $\delta E/\delta f(y)$ directly by using the fact that

$$E(f) = \int dx_1 \, dx_2 \, f(x_1)f(x_2), \qquad (1.81)$$

which clearly gives $\delta E/\delta f(y)$, agreeing with 1.80. We now draw attention to the fact that if we consider the functional $F = \int dx \, f(x)$, then we have just shown that

$$\begin{aligned}\frac{\delta F}{\delta f(y)} &= 1 \\ \frac{\delta F^2}{\delta f(y)} &= 2F.\end{aligned} \qquad (1.82)$$

It is easy to see that

$$\frac{\delta F^n}{\delta f(y)} = nF^{n-1}. \qquad (1.83)$$

The content of 1.82, 83 is that the functional F plays the role of the variable x in ordinary differential calculus. This prompts us to consider the analogue of ordinary power series in the variable x like, for example, the well-known Taylor series. A functional power series is not quite a series in powers of F but it may be constructed as follows: let

$$P(f) = K_0(x) + \int K_1(x, x_1)f(x_1)\,dx_1 \\ + \int K_2(x, x_1, x_2)f(x_1)f(x_2)\,dx_1\,dx_2 + \cdots \quad (1.84)$$

then $P(f)$ is called a functional power series in f. Now notice that

$$P(0) = K_0(x)$$

and

$$\frac{\delta P(f)}{\delta f(y)} = K_1(x, y) \qquad f = 0 \quad (1.85)$$

but that

$$\frac{\delta^2 P(f)}{\delta f(y_1)\delta f(y_2)} = K_2(x, y_1, y_2) + K_2(x, y_2, y_1) \qquad f = 0. \quad (1.86)$$

This last equation means that the coefficients $K_n(x, x_1 \cdots x_n)$ do not quite have the interpretation that we might have expected, namely that

$$n!\,K_n(x, x_1, \ldots, x_n) \quad \text{is} \quad \frac{\delta^n P(f)}{\delta f(x_1) \cdots \delta f(x_n)} \quad \text{at } f = 0.$$

However all is not lost; if $K_2(x, y_1, y_2)$ is symmetric in y_1 and y_2 then eq. 1.86 has the desired connection between coefficient function and functional derivative. This means that the general statement is that provided $K_n(x, y_1 \cdots x_n)$ is symmetric under permutation of the variables $x_1 \cdots x_n$ then

$$n!\,K_n(x, x_1 \ldots x_n) = \frac{\delta^n P(f)}{\delta f(x_1) \ldots \delta f(x_n)}, \qquad f = 0. \quad (1.87)$$

In our examples in field theory we shall consider functional power series which have this desirable symmetry property. Also when Fermions are considered we shall consider $K_n(x, x_1 \ldots x_n)$ where $K_n(x, x_1 \ldots x_n)$ is anti-symmetric under interchange of pairs of the variables $x_1 \ldots x_n$. Lest the reader think that this antisymmetry property would make $P(f)$ zero we add that in the Fermion case $f(x)$ will not be a simple function but an element of a Grassmann algebra. Let us now give a field theoretic example in which the coefficients $K_n(x, x_1 \ldots x_n)$ are symmetric in $x_1 \ldots x_n$. Consider the time ordered product of two scalar fields and take its vacuum expectation value (VEV)

$$\langle 0|T(\phi(x)\phi(y))|0\rangle = \theta(x_0 - y_0)\langle 0|\phi(x)\phi(y)|0\rangle$$
$$+ \theta(y_0 - x_0)\langle 0|\phi(y)\phi(x)|0\rangle. \quad (1.88)$$

We see straight away that $\langle 0|T(\phi(x)\phi(y))|0\rangle$ is symmetric in x and y. Further the VEV of the time ordered product of n scalar fields $\langle 0|T(\phi(x_1)\ldots\phi(x_n))|0\rangle$ is symmetric under interchange of pairs of the coordinates $x_1 \ldots x_n$. Hence if we define

$$K_n(x_1, \ldots x_n) = \frac{\langle 0|T(\phi(x_1)\ldots\phi(x_n))|0\rangle}{n!} \quad (1.89)$$

then the corresponding functional power series $P(f)$ has the property that we expect of a normal Taylor series in a real or complex variable namely

$$\frac{\delta^n P(f)}{\delta f(x_1) \ldots \delta f(x_n)} = \langle 0|T(\phi(x_1)\ldots\phi(x_n))|0\rangle \quad f = 0 \quad (1.90)$$

where

$$P(f) = \sum_0^\infty \int \frac{\langle 0|T(\phi(x_1)\ldots\phi(x_n))|0\rangle}{n!} f(x_1)\ldots f(x_n)\, dx_1 \ldots dx_n. \quad (1.91)$$

The particular power series $P(f)$ given in 1.91 plays an important part in field theory as we shall see later. It should also be remarked that in evaluating multiple derivatives like $\delta^n P(f)/\delta f(x_1) \ldots \delta f(x_n)$ the rules about differentiating a sum and a product of two functionals coincide with those of ordinary partial differential calculus. This should already be clear from the derivatives that we have already calculated.

Functionals and Fermions

We have alluded above to a possibility of considering functionals which are not operators on a space of functions but on a Grassmann algebra. This is the topic with which we are now going to deal. The first notion that must be dealt with and made precise is that of a

Grassmann algebra. A crude description of the elements of a Grassmann algebra is that they are 'anti-commuting C numbers'. Indeed once one has assimilated the definitions and the rules of Grassmann algebras the utility of that description becomes quite significant. A Grassmann algebra is an algebra with generators C_i obeying the equation shown below

$$\{C_i, C_j\} = C_i C_j + C_j C_i = 0. \tag{1.92}$$

If the index i runs from 1 to n the algebra is finite-dimensional of dimension 2^n. In our field theoretic calculations we shall deal with Grassmann algebras with an infinite number of generators. We record below some useful properties of a general Grassmann algebra. Equation 1.92 implies that for any element C_i

$$C_i^2 = 0, \tag{1.93}$$

also if $g(C)$ belongs to the algebra then $g(C)$ has an expansion in terms of the generators C_i

$$g(C) = G^0 + G^1_{i_1} C_{i_1} + G^2_{i_1 i_2} C_{i_1} C_{i_2} + \cdots G^n_{i_1 \ldots i_n} C_{i_1} \ldots C_{i_n} + \cdots \tag{1.94}$$

where the contracted indices $i_1 \ldots i_n$ are summed over the number of generators in the algebra. A useful property of a Grassmann algebra which has finite dimension equal to 2^n is that the expansion, 1.94, terminates after a finite number of terms. In fact the last term will be of the form $G^n_{i_1 \ldots i_n} C_{i_1} \ldots C_{i_n}$ so there are $n+1$ terms at most. The reason for this may be gleaned from 1.92, 93. The $(n+2)$nd term must vanish since it involves the product

$$C_{i_1} \ldots C_{i_n} C_{i_{n+1}}. \tag{1.95}$$

But since the dimension is 2^n there are only n distinct *generators* so at least two of the generators in 1.95 must be equal. Now we use the anti-commutation rule 1.92 to bring these generators adjacent to one another; then 1.93 shows that the resultant expression is zero. Clearly this is true for all subsequent terms in the expansion as well. This property has the intriguing consequence that expressions like

$$\exp[C(g)] \tag{1.96}$$

where $C(g)$ belongs to a finite-dimensional Grassmann algebra have only a finite number of terms in their power series expansion. This situation has already occurred in the literature in calculations in the superfield formalism for what are called supersymmetric Lagrangians.[3]

We now want to proceed to define derivatives of elements of Grassmann algebras. It turns out that because, unlike the case for ordinary functions, the elements of a Grassmann algebra do not commute, there are two sorts of derivative. They are called the left and right derivative and their properties only become important when differentiating products of the generators. The derivatives are linear in their operation and so the only requirement is to define the derivatives of generators and their products. The left derivative of a product $C_{i_1} \ldots C_{i_m}$ of generators is defined by

$$\frac{\partial}{\partial C_j} C_{i_1} \ldots C_{i_m} = \delta_{i_1 j} C_{i_2} \ldots C_{i_m} - \delta_{i_2 j} C_{i_1} C_{i_3} \ldots C_{i_m}$$
$$+ \cdots (-1)^{m-1} \delta_{i_m j} C_{i_1} \ldots C_{i_{m-1}}. \quad (1.97)$$

The right derivative is defined by

$$C_{i_1} \ldots C_{i_m} \frac{\partial}{\partial C_j} = C_{i_1} C_{i_2} \ldots C_{i_{m-1}} \delta_{i_m j} - C_{i_1} \ldots C_{i_{m-2}} C_{i_m} \delta_{i_{m-1} j}$$
$$+ \cdots (-1)^{m-1} C_{i_2} \ldots C_{i_m} \delta_{i_1 j}. \quad (1.98)$$

We can summarize these equations in words by saying; to compute the left derivative of $C_{i_s}, s = 1, \ldots, m$ with respect to C_j we commute C_{i_s} to the first place in the product $C_{i_1} \ldots C_{i_m}$ and then the derivative is simply C_{i_s} replaced by $\delta_{i_s j}$ with the appropriate power of -1. For the right derivative it is the last rather than the first place.

The notation that we shall use to denote the generators of an infinite-dimensional Grassmann algebra is $C(x)$ rather than C_i. The anti-commutation rule will then be written

$$\{C(x), C(x')\} = C(x)C(x') + C(x')C(x) = 0. \quad (1.99)$$

This means that we replace the relation $\partial C_i / \partial C_j = \delta_{ij}$ by

$$\frac{\delta C(x)}{\delta C(x')} = \delta(x - x'). \quad (1.100)$$

We are now ready to turn to functionals $F(C(x))$ where $C(x)$ belongs to an infinite-dimensional Grassmann algebra. Let us consider the power series

$$P(C(x)) = C_0(x) + \int C_1(x, x_1)C(x_1)\,dx_1$$
$$+ \cdots \int C_n(x, x_1 \ldots x_n)C(x_1)\ldots C(x_n)\,dx_1\ldots dx_n + \cdots. \quad (1.101)$$

We wish to establish the relation between the coefficient functions $C_n(x, x_1 \ldots x_n)$ and the derivatives of $P(C(x))$ at $C(x) = 0$. First of all we have

$$P(0) = C_0(x) \quad (1.102)$$

and

$$\frac{\delta P(0)}{\delta C(y_1)} = C_1(x, y_1). \quad (1.103)$$

But[†]

$$\frac{\delta}{\delta C(y_1)} \int dx_1\, dx_2\, C_2(x, x_1, x_2)C(x_1)C(x_2)$$
$$= \int dx_2\, C_2(x, y_1, x_2)C(x_2) - \int dx_1\, C_2(x, x_1, y_1)C(x_1). \quad (1.104)$$

Thus we can use 1.104 to establish that

$$\frac{\delta^2 P(0)}{\delta C(y_2)\delta C(y_1)} = C_2(x, y_1, y_2) - C_2(x, y_2, y_1). \quad (1.104a)$$

This clearly shows that it would be an advantage to have $C_2(x, x_1, x_2)$ antisymmetric in x_1 and x_2. Even more to the point if we look at the definition of $P(C(x))$ we see that the symmetric part of $C_2(x, x_1, x_2)$ contributes zero to the functional $P(C(x))$ because of the anti-commuting properties of $C(x_1)$ and $C(x_2)$. It is evident that analogous results hold for $C_n(x, x_1 \ldots x_n)$ for $n > 2$. A field theory with Fermions $\psi(x)$ provides an important example of coefficient functions $C_n(x, x_1 \ldots x_n)$ with the desired antisymmetry properties. Recall that the definition of the VEV of the time ordered product of the Fermi fields is

$$\langle 0 | T(\psi(x)\psi(y)) | 0 \rangle = \theta(x_0 - y_0)\psi(x)\psi(y) - \theta(y_0 - x_0)\psi(y)\psi(x). \quad (1.105)$$

[†] We are using here the left rather than the right derivative.

The minus sign in 1.105 is crucial. The choice of the coefficients $C_n(x', x_1 \ldots x_n)$ should now be obvious it is

$$C_n(x, x_1 \ldots x_n) = \frac{\langle 0|T(\psi(x_1) \ldots \psi(x_n))|0\rangle}{n!}. \tag{1.106}$$

The definition 1.106 makes $C_n(x, x_1 \ldots x_n)$ antisymmetric with respect to interchange of pairs of the coordinates $x_1 \ldots x_n$. It also gives us the formula

$$\frac{\delta^n P(0)}{\delta C(x_n) \ldots \delta C(x_1)} = \langle 0|T(\psi(x_1) \ldots \psi(x_n))|0\rangle$$

$$= n! C_n(x_1, \ldots x_n). \tag{1.107}$$

Functional Taylor Series

We are now in a position to deal with the question of the functional Taylor series. In the theory of ordinary functions one is familiar with the fact of life that not every function has a Taylor expansion. The derivatives which form the coefficients in the series must at least exist and the series should then converge. The existence of a functional Taylor series for an arbitrary functional $F(f)$ is also not automatic. The matter may be cleared up by considering the functional $F(f + \lambda f')$ and an alternative method for defining the functional derivative $\delta F/\delta f(y)$. The definition for the derivative is

$$\int dy\, f'(y) \frac{\delta F(f)}{\delta f'(y)} = \lim_{\lambda \to 0} \frac{[F(f + \lambda f') - F(f)]}{\lambda}. \tag{1.108}$$

That is to say we define the quantity $\delta F(f)/\delta f'(y)$ smeared by the function $f'(y)$, rather than $\delta F(f)/\delta f'(y)$ directly. This smeared object is mathematically more well behaved than the derivative itself. It is a simple exercise to show that this definition gives the same result as the previous one. The reader is encouraged though to evaluate the derivatives of several functionals by both methods and verify that the results coincide. Next let us consider the ordinary function $g(\lambda)$ of the real variable λ defined by

$$g(\lambda) = F(f + \lambda f') \tag{1.109}$$

where F is a functional. Suppose that $g(\lambda)$ has a Taylor series

$$g(\lambda) = \sum_0^\infty \frac{g^{(n)}(0)}{n!} \lambda^n, \qquad (1.110)$$

then

$$g^{(1)}(0) = \lim_{\lambda \to 0} \frac{F(f+\lambda f') - F(f)}{\lambda}$$

$$= \int f'(y) \frac{\delta F(f)}{\delta f'(y)} dy. \qquad (1.111)$$

Induction may be used to prove that

$$g^{(n)}(0) = \int dx_1 \ldots dx_n f'(x_1) \ldots f'(x_n) \frac{\delta^n F(f)}{\delta f'(x_1) \ldots \delta f'(x_n)}. \qquad (1.112)$$

This means that the power series $g(\lambda)$ appearing in 1.110 may be written as

$$F(f+\lambda f') = \sum_0^\infty \int dx_1 \ldots dx_n f'(x_1) \ldots f'(x_n) \frac{\delta^n F(f)}{\delta f'(x_1) \ldots \delta f'(x_n)} \frac{\lambda^n}{n!}. \qquad (1.113)$$

Now suppose that this power series converges when $\lambda = 1$ and then set $\lambda = 1$ and $f = 0$ and dispense with the prime on the function $f'(x)$. The result is the desired functional Taylor series.

$$F(f) = \sum_0^\infty \frac{1}{n!} \int dx_1 \ldots dx_n f(x_1) \ldots f(x_n) \frac{\delta^n F(0)}{\delta f(x_1) \ldots \delta f(x_n)}. \qquad (1.114)$$

We can summarize all this. $F(f)$ has the functional Taylor series expansion given in 1.114 when the ordinary function $g(\lambda)$ defined by $g(\lambda) = F(f+\lambda f')$, has an ordinary Taylor series convergent at the point $\lambda = 1$.

Functional and Path Integrals

I would now like to complement the preceding discussion of differentiation applied to spaces of functions, or spaces of 'anti-commuting functions' (Grassmann algebras) by defining an integration on such spaces. We shall begin with the space of functions; let us assume that all these functions $f(x)$ are square integrable. Then we know that the space of such functions is the familiar Hilbert space. This is an infinite-dimensional linear space. The fact that the space is infinite-dimensional is important. The reason is that to integrate in an ordinary Euclidean space with a finite number n of dimensions requires a multiple integral with n separate integrations. This means that to integrate in a Hilbert space one would expect that a multiple integral requiring an infinite number of separate integrations would be needed. The preceding simple line of reasoning is indeed correct and it leads to the correct speculation that such integrals will not always converge, (and hence exist), and that they will be difficult to compute.

The first appearance of the method of functional integration in quantum theory was due to Feynman (his methods are presented with great clarity in reference (4)). In quantum theory the term path integral is often used interchangeably with the term functional integral. This is because the integrals considered by Feynman can be described as integrals over a space of functions each of which describes the path of a particle with position x from point a to point b. However functional integrals in the sense of integrals over a space of functions were considered much earlier by the mathematician Wiener.[5] This historical detail is not in itself of great interest. The interest lies in the fact that the functional integrals constructed by Wiener and those considered in quantum theory are not quite the same. This difference is worth understanding. In order to compare the two sorts of integrals consider first that due to Wiener. This had its origin in the theory of Brownian motion. That is to say a particle diffusing randomly so that its probability distribution is given by the diffusion equation

$$\frac{\partial \psi}{\partial t} = D \frac{\partial^2 \psi}{\partial x^2} \qquad (1.115)$$

with

$$\psi = \frac{1}{(4\pi Dt)^{1/2}} \exp\left(\frac{-x^2}{4Dt}\right). \qquad (1.116)$$

We would like to construct an averaging process over the positions of the particle. Let the position of the particle at time t be $x(t)$. Divide the time interval $[0, t]$ into n equal parts by the points $t_1, \ldots t_{n-1}, t_n$ ($t_n = t$). Let the position of the particle $x(t_i)$ at time t_i be in the interval $[a_i, a'_i]$ for $i = 1 \ldots n$. Then the probability of this distribution of positions is given by the n-fold multiple integral†

$$\frac{1}{[\pi^n t_1(t_2-t_1)\dots(t_n-t_{n-1})]^{1/2}} \int_{a_1}^{a_1'} \int_{a_2}^{a_2'} \dots$$
$$\cdot \int_{a_n}^{a_n'} \exp\left[-\frac{x_1^2}{t_1} - \frac{(x_2-x_1)^2}{(t_2-t_1)} \dots - \frac{(x_n-x_{n-1})^2}{(t_n-t_{n-1})}\right] dx_1 \dots dx_n. \quad (1.117)$$

We now pass to the limit where $\Delta t = (t/n) \to 0$ and we insert in the integrand of 1.117 a function $F(x_1,\dots x_n)$‡. Then we write

$$\int F(x(t))\, dW = \lim_{n\to\infty} \frac{1}{(\pi\,\Delta t)^{n/2}} \int \dots \int F(x_1 \dots x_n)$$
$$\cdot \exp\left[-\frac{x_1^2}{\Delta t} - \sum_{i=1}^{n-1} \frac{(x_{i+1}-x_i)^2}{\Delta t}\right] dx_1 \dots dx_n. \quad (1.118)$$

The symbol dW stands for Wiener measure. We rewrite the RHS of 1.118 as

$$\frac{1}{A} \int \dots \int F(x(t)) \exp\left[-\int_0^t \frac{dx(t)}{dt'} dt'\right] \prod_0^t dx(t). \quad (1.119)$$

This enables us to give a symbolic representation of Wiener measure as

$$dW = \frac{1}{A} \exp\left[-\int_0^t \left(\frac{dx(t')}{dt'}\right)^2 dt'\right] \prod_0^t dx(t). \quad (1.120)$$

The factor $1/A$ is included to normalize the measure so that

$$\int dW = 1. \quad (1.121)$$

To illustrate the working of this measure an example is needed. The functional that we choose is

$$F(x(t)) = \exp\left[\int_0^t x^2(t')\, dt'\right] \quad (1.122)$$

so that we have to evaluate

$$\int \exp\left[\int_0^t x^2(t')\, dt'\right] dW$$
$$= \frac{1}{A} \int \exp\left[\int_0^t x^2(t')\, dt'\right] \exp\left[-\int_0^t \left(\frac{dx(t')}{dt'}\right)^2 dt'\right] \prod_0^t dx(t). \quad (1.123)$$

† For algebraic convenience we have set $D = \frac{1}{4}$.
‡ The integral of $F(x(t))$ can only be defined in terms of this limiting procedure defined on $F(x_1,\dots x_n)$ if $F(x_1,\dots x_n)$ itself is suitably defined. For more information consult reference 8.

The method used for evaluating 1.123 relies on a judiciously chosen linear change of variable. Hence we must first consider the behaviour of a functional integral under a change of variable. A change of variable from the function $x(t)$ to $y(t)$ is expressed by the integral equation

$$y(t) = x(t) + \int_0^t K(t, t') x(t') \, dt'. \tag{1.124}$$

In changing the variables in 1.123 there are two parts of dW which we have to consider: the 'volume element' $\prod_0^t dx(t)$, and the exponential factor $\exp\{-\int_0^t [dx(t')/dt']^2 \, dt'\}$. The volume element $\prod_0^t dx(t)$ transforms under a kind of infinite dimensional generalization of a Jacobian. It becomes simply $D(t) \prod_0^t dx(t)$ where D is Fredholm's determinant. The Fredholm determinant can be constructed by replacing the integral in 1.124 by a sum in which the interval $[0, t]$ is split into n pieces. The integral equation then becomes a system of linear equations in n unknowns. The limit of the Jacobian determinant for this system of equations as $n \to \infty$ is the Fredholm determinant. In eq. 1.124 it may be assumed without loss of generality that $K(t, t') = 0$ for $t < t'$. Integral equations with this property are called Volterra equations. In this case the formula for the Fredholm determinant is simply

$$D(t) = \exp\left[-\tfrac{1}{2} \int_0^t K(t', t') \, dt'\right]. \tag{1.125}$$

The details of these properties of integral equations are readily found in Chapter 4 of reference 7.

The exponential factor in dW becomes

$$\exp\left[-\left[\int_0^t \frac{dx(t')}{dt'} + \frac{d}{dt'} \int_0^{t'} K(t', t'') x(t'') \, dt''\right]^2\right]$$

$$= \exp\left[\left[-\int_0^t \frac{dx(t')}{dt'}\right]^2\right] \exp\left[-2 \int_0^t \frac{d}{dt'} \int_0^{t'} K(t', t'') x(t'') \, dt'' \, dx(t')\right]$$

$$\cdot \exp\left[-\int_0^t \left\{\frac{d}{dt'} \int_0^{t'} K(t', t'') x(t'') \, dt''\right\}^2 dt'\right]$$

$$= \exp\left[-\int_0^t \left(\frac{dx(t')}{dt'}\right)^2\right] \exp\left[\int_0^t A(t') \, dt'\right]. \tag{1.126}$$

Thus the transformation property of dW is

$$dW \mapsto D(t) \exp\left[\int_0^t A(t') \, dt'\right] dW \tag{1.127}$$

Now the ingenious technique that is used to evaluate the integral 1.123 may be revealed. Start with the integral of 1.121 which we know to be unity and then change variables using 1.124. The result of this is the equation

$$\int dW = 1 = D \int \exp\left[\int_0^t A(t')\, dt'\right] dW \tag{1.128}$$

with $\exp\left[\int_0^t A(t')\, dt'\right]$ given by 1.126. We are now free to choose the kernel $K(t, t')$ so that

$$\exp\left[\int_0^t A(t')\, dt'\right] = \exp\left[\int_0^t x^2(t')\, dt'\right].$$

Hence by 1.128 we have that

$$\int \exp\left[\int_0^t x^2(t')\, dt'\right] dW = D^{-1}. \tag{1.129}$$

The choosing of $K(t, t')$ is simple once one uses the equation

$$\exp\left[-\int_0^t \left\{\frac{d}{dt'}\int_0^{t'} K(t', t'') x(t'')\, dt''\right\}^2 dt' \right.$$
$$\left. -2\int_0^t \frac{d}{dt'}\int_0^{t'} K(t', t'') x(t'')\, dt'\, dx(t')\right]$$
$$= \exp\left[-\int_0^t K^2(t', t') x^2(t')\, dt' + \int_0^t K(t', t')\, d(x^2(t'))\right]$$
$$= \exp\left[-\int_0^t K^2(t', t') x^2(t')\, dt' - \int_0^t \frac{dK(t', t')}{dt'} x^2(t')\, dt'\right]. \tag{1.130}$$

It is now clear that $\exp\left[\int_0^t A(t')\, dt'\right]$ will be equal to $\exp\left[\int_0^t x^2(t')\, dt'\right]$ provided $K(t, t')$ is chosen so that

$$\frac{dK(t', t')}{dt'} + K^2(t', t') = -1. \tag{1.131}$$

If we use the substitution $K(t', t') = dJ/dt \cdot 1/J$ then the differential equation (1.131) becomes

$$\frac{d^2 J}{dt'^2} + J = 0. \tag{1.132}$$

This has as solution

$$J(t') = A \cos t' + B \sin t'. \tag{1.133}$$

The constants A and B may be fixed by using the initial conditions.† The conditions are that $J(t'=t) = 1$ and $d/dt'\, J(t'=t) = 0$, thus $J(t')$ may be written

$$J(t') = \cos(t' - t). \tag{1.134}$$

Finally, using 1.125

$$\begin{aligned} D^{-1} &= \exp\left[\frac{1}{2}\int_0^t K(t',t')\,dt'\right] \\ &= \exp\left[\frac{1}{2}\int_0^t \frac{dJ(t')}{dt'}\frac{1}{J(t')}\,dt'\right] \\ &= \exp\left[\frac{1}{2}\ln\left(\frac{J(t)}{J(0)}\right)\right] \\ &= [(\cos t)^{1/2}]^{-1} \end{aligned} \tag{1.135}$$

The reason that the answer depends on t is, of course, that the original functional $\exp\left[\int_0^t x^2(t')\,dt'\right]$ depends on t. To write our result in full we have

$$\int \exp\left[\int_0^t x^2(t')\,dt'\right] dW = (\cos t)^{-1/2}. \tag{1.136}$$

If this evaluation of a functional integral has left the reader feeling a little perplexed or confused he should not worry. This is quite normal and we shall not in any case have to evaluate any more such integrals in this sort of detail in the rest of this book.

We now want to describe the integrals introduced by Feynman. The path integral of Feynman was introduced to evaluate the sum of the contribution of all the possible paths that a particle can take in moving from a to b. Of course, we know that in classical mechanics only *one* path is important, namely that for which the action is an extremum. In quantum mechanics this is not quite right. Every path is now important; it is just that the path which makes the action attain its extremal value, or paths which give a value within $\hbar$ of the extremal

---

† It requires a little more work to see that these are the correct conditions. We leave this out here for pedagogical reasons; see however reference 6.

value, are the most important. In fact if $S(x(t))$ is the action for a path $x(t)$ then its amplitude is known[4] to be given by

$$\exp\left[\frac{i}{\hbar}S(x(t'))\right]. \tag{1.137}$$

Then the integral over all paths is given by

$$I = \int \exp\left[\frac{i}{\hbar}S(x(t))\right]\Pi\, dx(t). \tag{1.138}$$

with $\Pi\, dx(t)$ as before. The action S is the time integral of the Lagrangian L where

$$L = \left(\frac{dx}{dt}\right)^2 - V(x). \tag{1.139}$$

Then the integral I is simply

$$\int \exp\left[-\frac{i}{\hbar}\int_0^t V(x)\, dt'\right]\exp\left[\frac{i}{\hbar}\int_0^t \left(\frac{dx(t')}{dt'}\right)^2 dt'\right]\Pi\, dx(t). \tag{1.140}$$

Equation 1.140 would appear to be the integral over Wiener measure of the quantity $\exp[-i/\hbar \int_0^t V(x)\, dt']$. However this is not in fact so because of the presence of the factor $i/\hbar$ in the exponential. The factor $\hbar$ is not important; its presence is simply a matter of units. But the factor i is a fundamental change. This factor corresponds to a diffusion with diffusion coefficient D which is imaginary. That is to say the diffusion equation 1.115 is replaced by,

$$\frac{i}{\hbar}\frac{\partial\psi}{\partial t} = D\frac{\partial^2\psi}{\partial x^2} \quad (V = 0). \tag{1.141}$$

This equation is recognized straightaway as the Schrodinger equation. Since the equation lies at the basis of quantum theory we see the futility of trying to tamper with 1.140. We must accept it as it stands. The presence of the i in the exponential means that it causes uncontrollable oscillations in the path integral. Despite this drawback it is a matter of history that it has not hindered the path integral from being an extremely important contribution to quantum theory. Hence we shall use the integral 1.138 in what follows and manipulate it formally as if it were a well-defined mathematical object. Having spent some time explaining the ideas that provide the basis for functional integrals we shall now proceed in a much more formal manner. We wish to make use of functional integral techniques in field theory. The integral I of 1.138 will serve as our starting point. The action is taken to be the action of $\lambda\phi^4/4!$ theory

$$S = \int d^4x \left(\tfrac{1}{2}\partial_\mu\phi\, \partial^\mu\phi - \frac{m^2\phi^2}{2} - \frac{\lambda\phi^4}{4!} \right) \qquad (1.142)$$

and instead of the volume element $\Pi\, dx(t)$ of the space of functions we substitute a volume element $\Pi\, d\phi(x)$ defined on the space of classical fields. The integral then becomes, (with $\hbar = 1$)

$$\int \exp\left[i\int d^4x \left(\tfrac{1}{2}\partial_\mu\phi\, \partial^\mu\phi - \frac{m^2\phi^2}{2} - \frac{\lambda\phi^4}{4!} \right) \right]. \qquad (1.143)$$

The volume element $\Pi\, d\phi(x)$ is usually written $\mathscr{D}[\phi]$. A quantum field theory can be considered to be characterized by the set of all its Green's functions $\langle 0 | T(\phi(x_1) \ldots \phi(x_n)) | 0 \rangle$. These Green's functions can all be written compactly using functional integrals. This is accomplished by using a very useful technique due to Schwinger. The technique consists of modifying the action 1.142 by the addition of the term $\int d^4x J(x)\phi(x)$ where $J(x)$ is not an operator but an ordinary scalar function. The exponential of the action, i.e. the integrand of the functional integral is then written as

$$\exp\left[i\int d^4x \left(\tfrac{1}{2}\partial_\mu\phi\partial^\mu\phi - \frac{m^2\phi^2}{2} - \frac{\lambda\phi^4}{4!} + J(x)\phi(x) \right) \right]. \qquad (1.144)$$

The resulting functional integral we write as $F[J]$, since it is a functional of the source $J(x)$, where

$$F[J] = \int \mathscr{D}[\phi] \exp\left[i\int dx \left(\tfrac{1}{2}\partial_\mu\phi\partial^\mu\phi - \frac{m^2\phi^2}{2} - \frac{\lambda\phi^4}{4!} + J\phi \right) \right]. \qquad (1.145)$$

The great utility of the functional $F[J]$ is that the moments with respect to ϕ of the integral 1.145, which we shall see soon are just the Green's functions, are given by simply functionally differentiating $F[J]$ with respect to $J(x)$. For example

$$\frac{\delta F[J]}{\delta J(x_1)} = i\int \mathscr{D}[\phi]\phi(x_1) \exp\left[i\int dx \left(\tfrac{1}{2}\partial_\mu\phi\partial^\mu\phi - \frac{m^2\phi^2}{2} - \frac{\lambda\phi^4}{4!} + J\phi \right) \right].$$
$$(1.146)$$

An nth order moment is given by

$$(-i)^n \frac{\delta^n F[J]}{\delta J(x_n) \ldots \delta J(x_1)} = \int \mathscr{D}[\phi] \phi(x_1) \ldots \phi(x_n)$$
$$\cdot \exp\left[i \int dx \left(\tfrac{1}{2}\partial_\mu \phi \partial^\mu \phi - \frac{m^2 \phi^2}{2} - \frac{\lambda \phi^4}{4!} + J\phi\right)\right].$$
(1.147)

We now want to calculate some field-theoretic quantities by doing a functional integral over the classical fields. The functional integral that we shall attempt is $F[J]$ itself. It will have been noticed by the observant reader that our success in performing the functional integral 1.136 arises from the integrand's being the exponential of a quadratic form. In fact it is not known how to evaluate directly functional integrals where the quadratic form is replaced by cubic, quartic terms etc. Therefore we shall not be able to deal with the $\lambda \phi^4/4!$ term in 1.145. Instead we shall compute $F[J]$ when $\lambda = 0$ and show that $F[J]$ and its functional derivatives $(-i)^n [\delta^n F/\delta J(x_1) \ldots J(x_n)]$ give the free particle Green's functions. The labour of computing $F[J]$ consists entirely of manipulations designed to reduce the argument of the exponential to a quadratic form. The first term to require attention is $(\partial_\mu \phi)^2/2$. Simple differentiation shows us that

$$\partial_\mu (\partial_\mu \phi^2) = 2(\phi \Box \phi + \partial_\mu \phi \partial^\mu \phi),$$
(1.148)

thus

$$\tfrac{1}{2}(\partial_\mu \phi)^2 = -\tfrac{1}{2}\phi \Box \phi + \tfrac{1}{4}\partial_\mu (\partial_\mu \phi^2).$$
(1.149)

But the terms $\tfrac{1}{2}(\partial_\mu \phi)^2$ and $-\tfrac{1}{2}\phi \Box \phi$ differ only by a total derivative; it is a familiar fact that total derivatives contribute zero to the action integral.† We shall therefore replace $\tfrac{1}{2}(\partial_\mu \phi)^2$ by $-\tfrac{1}{2}\phi \Box \phi$. The kinetic terms are now of the form $\int \phi(\Box + m^2)\phi(x)$. Our next task is to absorb the differential operator $\Box + m^2$ by redefining the field $\phi(x)$. This requires a few steps. The algebraic structure of what we are doing is simple to describe. We regard the kinetic term as being of the form $\phi A \phi$. Then, if we define $\phi' = A^{1/2}\phi$, the resultant term is of the form $\phi'\phi'$. This task is made difficult by the fact that A is a differential operator and we do not know how to take its square root. We therefore pass to momentum space and define a function $\hat{K}(p)$ by the equation

$$(p^2 - m^2)^{1/2} \hat{K}(p) = 1$$
(1.150)

† We assume that the function vanishes at the end points.

or
$$\hat{K}(p) = (p^2 - m^2)^{-1/2}. \tag{1.151}$$

Then we introduce the Fourier transform $K(x)$ of $\hat{K}(p)$. Evidently

$$K(x) = \frac{1}{(2\pi)^4} \int dp \, \frac{e^{ipx}}{(p^2 - m^2)^{1/2}} \tag{1.152}$$

The task of taking the square root of $\Box + m^2$ has been accomplished indirectly by Fourier transformation. The change of field variable that we require is now clear; it is

$$\phi(x) = \int dy \, K(x-y) \phi'(y). \tag{1.153}$$

The kinetic term in the exponential may now be written as

$$\int dx \, dx_1 \, dx_2 \, \phi'(x_1) K(x-x_1)(\Box + m^2) K(x-x_2) \phi'(x_2) = \int dx \, F(x) G(x), \tag{1.154}$$

where
$$F(x) = \int dx_1 \, K(x-x_1) \phi'(x_1)$$

and
$$G(x) = \int dx_2 \, (\Box + m^2) K(x-x_2) \phi'(x_2). \tag{1.155}$$

But it is a well known property of Fourier transforms that if $\hat{F}(p)$ and $\hat{G}(p)$ are the Fourier transform of $F(x)$ and $G(x)$ respectively then

$$\int dx \, F(x) G(x) = \int dp \, \hat{F}(p) \hat{G}(p) \tag{1.156}$$

but
$$\hat{F}(p) = (p^2 - m^2)^{-1/2} \hat{\phi}'(p) \tag{1.157}$$

since $F(x)$ is just a convolution. Similarly we may evaluate $\hat{G}(p)$ to obtain
$$\hat{G}(p) = -(p^2 - m^2)(p^2 - m^2)^{-1/2} \hat{\phi}'(p)$$
$$= -(p^2 - m^2)^{1/2} \hat{\phi}'(p). \tag{1.158}$$

Bringing these results together gives

$$\int dx \, dx_1 \, dx_2 \, \phi'(x_1) K(x-x_1)(\Box + m^2) K(x-x_2) \phi'(x_2)$$
$$= \int dx \, F(x) G(x) = -\int dp \, \hat{\phi}(p) \hat{\phi}(p) = -\int dx . \phi'(x) \phi'(x). \tag{1.159}$$

We have now achieved the form that we required and can now change the variable in the functional integral from ϕ to ϕ'. This means that

$$F[J] = D \int \mathcal{D}[\phi] \exp\left[i \int dx \{\tfrac{1}{2}\phi(x)\phi(x) + \int dx_1 J(x)K(x-x_1)\phi(x_1)\}\right] \quad (1.160)$$

where D is the Fredholm determinant for K. The last term to be dealt with is the source term. It is straightforward to find a change of variable to eliminate the source term. The choice needed is

$$\phi(x) = \phi'(x) - \int dx_1 J(x)K(x-x_1). \quad (1.161)$$

(Since it is only a constant function that is added on to $\phi(x)$ the Fredholm determinant in this case is unity.) With this change of variable the argument of the exponential becomes

$$\int dx \left[\frac{\phi'(x)\phi'(x)}{2} - \int dx\, dx_1 J(x)K(x-x_1)\phi(x_1) \right.$$
$$+ \frac{1}{2}\int dx\, dx_1\, dx_2 J(x)K(x-x_1)J(x)K(x-x_2)$$
$$\left. - \int dx\, dx_1 J(x)K(x-x_1)\phi(x_1) \right]. \quad (1.162)$$

[In the second term the property $K(x-x_1) = K(x_1-x)$ was used; this follows from its definition (1.152)]. The second and last terms cancel. The third term is again of the form $\tfrac{1}{2}\int dx\, H(x)H(x)$ with $H(x) = \int dx_1 J(x)K(x-x_1)$. We deal with this term in the same way as we did above.

$$\tfrac{1}{2}\int dx\, H(x)H(x) = \tfrac{1}{2}\int dp\, \hat{H}(p)\hat{H}(p)$$
$$= \tfrac{1}{2}\int dp\, \hat{J}(p)$$
$$\cdot (p^2-m^2)^{-1/2}(p^2-m^2)^{-1/2}\hat{J}(p)$$
$$= \tfrac{1}{2}\int dp\, \frac{\hat{J}(p)\hat{J}(p)}{p^2-m^2}$$
$$= \tfrac{1}{2}\int dp\, \hat{J}(p)\hat{E}(p) \quad (1.163)$$

where
$$\hat{E}(p) = \frac{\hat{J}(p)}{p^2 - m^2}.$$

Therefore we have

$$\frac{1}{2}\int dx\, H(x)H(x) = \frac{1}{2}\int dx\, J(x)E(x)$$
$$= -\frac{1}{2}\int dx\, dy\, J(x)\,\Delta_F(x-y)J(y) \quad (1.164)$$

where $\Delta_F(x-y)$ is the usual scalar particle Green's function satisfying

$$(\Box + m^2)\,\Delta_F(x-y) = \delta(x-y). \quad (1.165)$$

The functional $F[J]$ is now seen to be

$$F[J] = \exp\left[-\frac{i}{2}\int dx\, dy\, J(x)\,\Delta_F(x-y)J(y)\right]$$
$$\cdot D\int \mathcal{D}[\phi]\exp\left[i\int \frac{\phi^2}{2}dx\right]. \quad (1.166)$$

The factor $D\int \mathcal{D}\phi \exp[\frac{1}{2}\int \phi^2\, dx]$ does not depend on J and is simply a number. We eliminate it by redefining the integral that we are interested in as

$$F[J] = \frac{\int \mathcal{D}[\phi]\exp\left[i\int dx\left\{\frac{\partial_\mu\phi\partial^\mu\phi}{2} - \frac{m^2\phi^2}{2} + J\phi\right\}\right]}{\int \mathcal{D}[\phi]\exp\left[i\int dx\left\{\frac{\partial_\mu\phi\partial^\mu\phi}{2} - \frac{m^2\phi^2}{2}\right\}\right]}$$
$$= \exp\left[-\frac{i}{2}\int dx\, dy\, J(x)\,\Delta_F(x-y)J(y)\right]. \quad (1.167)$$

Note $F[0] = 1$. The functional $F[J]$ has a functional Taylor series obtained by expanding the exponential. The terms in this Taylor series turn out to be the free particle Green's functions. For example we have

$$\left.\frac{\delta^2 F[J]}{\delta J(x_1)\,\delta J(x_2)}\right|_{J=0} = -i\,\Delta_F(x_1 - x_2). \quad (1.168)$$

But since $\Delta_F(x_1 - x_2)$ satisfies 1.165, then $i\,\delta^2 F[0]/[\delta J(x_1)\,\delta J(x_2)]$ is the propagator of a scalar F particle of mass m. Similarly we can compute that

$$\frac{\delta^4 F[J]}{\delta J(x_1)\,\delta J(x_2)\,\delta J(x_3)\,\delta J(x_4)}\bigg|_{J=0} = \frac{\delta^4 F[0]}{\delta J(x_1)\,\delta J(x_2)\,\delta J(x_3)\,\delta J(x_4)}$$

$$= -\Delta_F(x_1 - x_2)\,\Delta_F(x_3 - x_4) - \Delta_F(x_1 - x_3)\,\Delta_F(x_2 - x_4)$$

$$- \Delta_F(x_1 - x_4)\,\Delta_F(x_2 - x_3). \tag{1.169}$$

We can represent 1.168,169 diagramatically below in Figs 10(a) and (b). Figure 10(a) simply depicts the propagation of the particle from the space-time point x_1 to the space-time point x_2. Figure 10(b) depicts

$$\frac{\delta^2 F[0]}{\delta J(x_1)\,\delta J(x_2)} = x_1 \text{———} x_2$$

FIG. 10(a)

$$\frac{\delta^4 F[0]}{\delta J(x_1)\delta J(x_2)\delta J(x_3)\delta J(x_4)} = \begin{matrix} x_1 \text{———} x_2 \\ x_3 \text{———} x_4 \end{matrix} + \begin{matrix} x_1 & x_2 \\ | & | \\ x_3 & x_4 \end{matrix} + \begin{matrix} x_1 \quad\ \ x_2 \\ \diagdown\!\diagup \\ \diagup\!\diagdown \\ x_3 \quad\ \ x_4 \end{matrix}$$

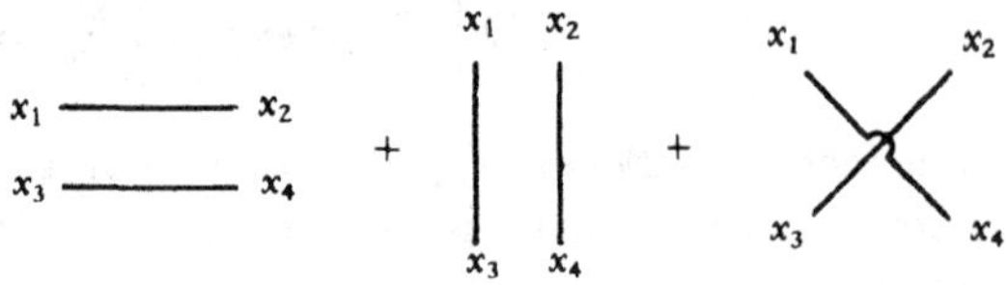

FIG. 10(b)

the possible motions of two free particles from the pair of points x_1, x_3 to the pair of points x_2, x_4. It is obvious that the higher derivatives $\delta^n F[0]/[\delta J(x_1) \ldots \delta J(x_n)]$ vanish for odd n; and for even n are given by sums of products of the function $\Delta_F(x - y)$. This means that $\delta^n F[0]/[\delta J(x_1) \ldots \delta J(x_n)]$ are simply the free particle Green's functions, i.e. the quantities $\langle 0|T(\phi(x_1) \ldots \phi(x_n))|0\rangle$ where $\phi(x)$ is a free scalar field. The evaluation of the Green's functions when given in terms of the vacuum expectation value of a time-ordered product is done by repeated use of the relation for the product of two free fields

$$\langle 0|\phi(x)\phi(y)|0\rangle = i\,\Delta^+(y - x) \tag{1.170}$$

where $\Delta^+(x - y)$ satisfies the equation

$$(\Box + m^2)\,\Delta^+ = 0. \tag{1.171}$$

RENORMALIZATION AND FUNCTIONAL DIFFERENTIATION 37

To summarize the situation for free fields we have that

$$(-i)^n \frac{\delta^n F[0]}{\delta J(x_1) \ldots \delta J(x_n)} = \langle 0 | T(\phi(x_1) \ldots \phi(x_n)) | 0 \rangle$$

$$= \frac{\int \mathscr{D}[\phi] \phi(x_1) \ldots \phi(x_n) \exp\left[i \int dx \left\{ \frac{(\partial_\mu \phi)^2}{2} - \frac{m^2 \phi^2}{2} \right\}\right]}{\int \mathscr{D}[\phi] \exp\left[i \int dx \left\{ \frac{(\partial_\mu \phi)^2}{2} - \frac{m^2 \phi^2}{2} \right\}\right]}. \quad (1.172)$$

The Green's functions are simply given by the moments of the function $\exp[i S(\phi)]$ when integrated functionally over ϕ. The great importance of this property for free fields is that it remains true for interacting fields. The only change being that the free action

$$\int dx \left\{ \tfrac{1}{2}(d_\mu \phi)^2 - \frac{m^2 \phi^2}{2} \right\}$$

is replaced by the complete action

$$\int dx \left\{ \tfrac{1}{2}(\partial_\mu \phi)^2 - \frac{m^2 \phi^2}{2} - \frac{\lambda \phi^4}{4!} \right\}.$$

We shall prove this last statement by showing that an expansion of $\exp[i S(\phi)]$ in powers of λ generates the Feynman graphs of $\lambda \phi^4/4!$ theory. For $\lambda \neq 0$ we rename $F[J]$ and define instead $Z[J]$ by

$$Z[J] = \frac{\int \mathscr{D}[\phi] \exp\left[i \int dx \left\{ \frac{(\partial_\mu \phi)^2}{2} - \frac{m^2 \phi^2}{2} - \frac{\lambda \phi^4}{4!} + J\phi \right\}\right]}{\int \mathscr{D}[\phi] \exp\left[i \int dx \left\{ \frac{(\partial_\mu \phi)^2}{2} - \frac{m^2 \phi^2}{2} - \frac{\lambda \phi^4}{4!} \right\}\right]}.$$

(1.173)

The presence of the quartic term $\lambda \phi^4/4!$ in the action means that we cannot perform the functional integral directly. But if we expand the numerator and denominator in powers of λ then the functional integrals become sums of moments of the free action. These integrals we have just learnt how to perform. In fact this expansion coincides with the usual perturbation expansion in terms of Feynman graphs. The best way to prove this is to calculate the term of order λ for a particular Green's function and show that it coincides with the appropriate Feynman graph of order λ. In following the steps in this calculation the means of extending this argument to a general one for all Green's functions calculated to all orders in λ becomes clear. The Green's function that we choose to evaluate to order λ is the four point function $\langle 0|T(\phi(x_1)\phi(x_2)\phi(x_3)\phi(x_4))|0\rangle$. In order that at the end of the calculation we can simply compare our answer with the perturbation theoretic one some remarks are helpful. In perturbation theory we usually work in momentum space and we always deal with connected diagrams. Connected Green's functions are defined by subtracting out those contributions to the scattering process in which some particles remain "spectators", i.e. do not scatter. We illustrate this for the four point function in Fig. 11. In Fig. 11 the left-hand diagram stands for

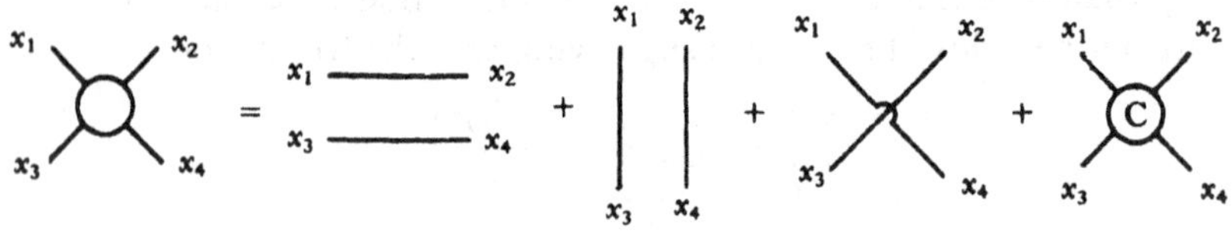

FIG. 11

the complete Green's function $\langle 0|T(\phi(x_1)\phi(x_2)\phi(x_3)\phi(x_4))|0\rangle$. The diagram on the RHS with the letter C stands for the connected Green's function which we denote by $\langle 0|T(\phi(x_1)\phi(x_2)\phi(x_3)\phi(x_4))|0\rangle_C$. The representation in Fig. 11 may also be written as the equation

$$\langle 0|T(\phi(x_1)\phi(x_2)\phi(x_3)\phi(x_4))|0\rangle$$
$$= \langle 0|T(\phi(x_1)\phi(x_2))|0\rangle \langle 0|T(\phi(x_3)\phi(x_4))|0\rangle$$
$$+ \langle 0|T(\phi(x_1)\phi(x_3))|0\rangle \langle 0|T(\phi(x_2)\phi(x_4))|0\rangle$$
$$+ \langle 0|T(\phi(x_1)\phi(x_4))|0\rangle \langle 0|T(\phi(x_2)\phi(x_3))|0\rangle$$
$$+ \langle 0|T(\phi(x_1)\phi(x_2)\phi(x_3)\phi(x_4))|0\rangle_C. \qquad (1.174)$$

After extracting the connected part from the full Green's function it is also the practice in perturbation theory to 'amputate' the external lines, i.e. remove the propagators attached to them. To denote an amputated Green's function of this sort we place a bar over the letter C. The amputated connected Green's function has then the definition

$$\langle 0|T(\phi(x_1)\phi(x_2)\phi(x_3)\phi(x_4))|0\rangle_{\bar{C}}$$
$$= \int d\xi_1\, d\xi_2\, d\xi_3\, d\xi_4\, \Delta_F^{-1}(x_1-\xi_1)\, \Delta_F^{-1}(x_2-\xi_2)\, \Delta_F^{-1}(x_3-\xi_3)$$
$$\cdot \Delta_F^{-1}(x_4-\xi_4) \langle 0|T(\phi(\xi_1)\phi(\xi_2)\phi(\xi_3)\phi(\xi_4))|0\rangle \qquad (1.175)$$

where the inverse propagator satisfies

$$\int d\xi \, \Delta_F^{-1}(x-\xi) \, \Delta_F(\xi-y) = \delta(x-y) \tag{1.176}$$

in coordinate space and

$$\hat{\Delta}_F^{-1}(p) \, \Delta_F(p) = 1 \tag{1.177}$$

in momentum space. Finally in passing to momentum space the Green's function $\langle 0|T(\phi(x_1)\phi(x_2)\phi(x_3)\phi(x_4))|0\rangle_{\bar{c}}$ depends, because of translational invariance, on only three variables, the coordinate differences x_1-x_2, x_1-x_3, x_1-x_4. This means that an energy-momentum conserving delta function always appears in the Fourier transform. The contribution to the series of Feynman diagrams in momentum space we write as $T(p_1 \ldots p_4)$ where $T(p_1 \ldots p_4)$ is defined by

$$\int dx_1 \ldots dx_4 \, e^{-ip_1 x_1} \ldots e^{-ip_4 x_4} \langle 0|T(\phi(x_1)\phi(x_2)\phi(x_3)\phi(x_4))|0\rangle_{\bar{c}}$$
$$= T(p_1 \ldots p_4)(-i)(2\pi)^4 \delta(p_1+p_2+p_3+p_4). \tag{1.178}$$

We have now, by our definitions, pinned down the precise object which the Feynman diagrams allow us to evaluate. For the four-point function there is only one Feynman diagram of order λ. After this

FIG. 12 The expression $-i\lambda$.

preamble of definitions we can now proceed to calculate $\langle 0|T(\phi(x_1)\phi(x_2)\phi(x_3)\phi(x_4))|0\rangle$ from which we shall proceed via our definitions to $T(p_1 \ldots p_4)$. Now consider $Z[J]$. If we expand numerator and denominator to order λ we obtain

$$\frac{\int \mathscr{D}[\phi] \exp\left[i \int dx \left\{\frac{(\partial_\mu \phi)^2}{2} - \frac{m^2 \phi^2}{2} + J\phi\right\}\right]\left[1 - \int dx \frac{\lambda}{4!}\phi^4(x)\right]}{\int \mathscr{D}[\phi] \exp\left[i \int dx \left\{\frac{(\partial_\mu \phi)^2}{2} - \frac{m^2 \phi^2}{2}\right\}\right]\left[1 - \int dx \frac{\lambda \phi^4}{4!}(x)\right]}.$$

$$\tag{1.179}$$

The denominator is of the form $(a - x)^{-1}$ so that expanding this to one order in λ immediately yields the order λ contribution to 1.179:

$$\left[\lambda \int \mathcal{D}[\phi] \exp\left[i \int dx \left\{\frac{(\partial_\mu \phi)^2}{2} - \frac{m^2 \phi^2}{2} + J\phi\right\}\right]\right.$$

$$\cdot \int \mathcal{D}[\phi] \int dx \frac{\phi^4}{4!}(x) \exp\left[i \int dx \left\{\frac{(\partial_\mu \phi)^2}{2} - \frac{m^2 \phi^2}{2}\right\}\right]$$

$$- \lambda \int \mathcal{D}[\phi] \int dx \frac{\phi^4}{4!}(x) \exp\left[i \int dx \left\{\frac{(\partial_\mu \phi)^2}{2} - \frac{m^2 \phi^2}{2} + J\phi\right\}\right]$$

$$\left[\int \mathcal{D}[\phi] \exp\left[i \int dx \left\{\frac{(\partial_\mu \phi)^2}{2} - \frac{m^2 \phi^2}{2}\right\}\right]\right]^{-1} = Z^{(1)}[J]. \quad (1.180)$$

To obtain the order λ contribution to the four point function one must simply evaluate

$$\frac{(-i)^4 \delta^4 Z^{(1)}[J]}{\delta J(x_1) \delta J(x_2) \delta J(x_3) \delta J(x_4)}$$

and set $J = 0$. But perusal of 1.180 shows quickly that this quantity may be written entirely in terms of the known functional $F[J]$. In fact we find that

$$\frac{(-i)^4 \delta^4 Z^{(1)}[0]}{\delta J(x_1) \delta J(x_2) \delta J(x_3) \delta J(x_4)} = \frac{\lambda}{4!} \frac{\delta^4 F[0]}{\delta J(x_1) \delta J(x_2) \delta J(x_3) \delta J(x_4)}$$

$$\cdot \int dx \frac{\delta^4 F[0]}{\delta J(y_1) \delta J(y_2) \delta J(y_3) \delta J(y_4)} - \frac{\lambda}{4!}$$

$$\int dx \frac{\delta^8 F[0]}{\delta J(x_1) \delta J(x_2) \delta J(x_3) \delta J(x_4) \delta J(y_1) \delta J(y_2) \delta J(y_3) \delta J(y_4)}$$

$$(1.181)$$

where

$$y_1 = y_2 = y_3 = y_4 = x, \quad J = 0. \quad (1.182)$$

Now because of the simple form of $F[J]$ the RHS of 1.181 is simply the sum of products of various numbers of functions $\Delta_F(x - y)$. Some of the Δ_F's will have their arguments integrated over because of the integral over x. The most sensible thing to do is to compute this expression directly by differentiation, and then represent the result compactly by using diagrams. The first term in 1.181 can be represented by the diagrams of Fig. 13. The second term in 1.181 is represented by the diagrams of Fig. 14. If we now add the contributions of Figs 13 and 14 we find that the diagrams of Fig. 13 cancel out

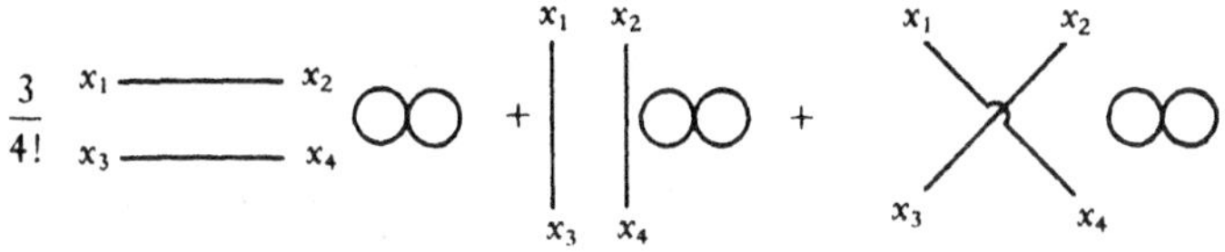

completely against the first three of Fig. 14. We now explain the diagrammatic notation. A line with coordinate indices x, y at either end represents the function $\Delta_F(x-y)$. A vertex stands for $\lambda \int dx$. A

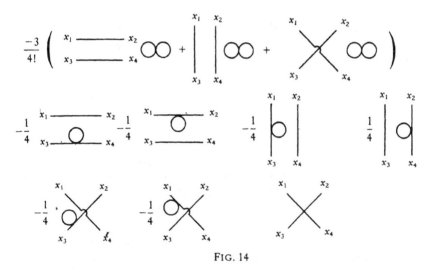

FIG. 14

bubble with no external lines is to be thought of as a line leaving a point x' and subsequently returning to it. It thus stands for $\Delta_F(x'-x') = \Delta_F(0)$. For example the two diagrams (Fig. 15) stand for the

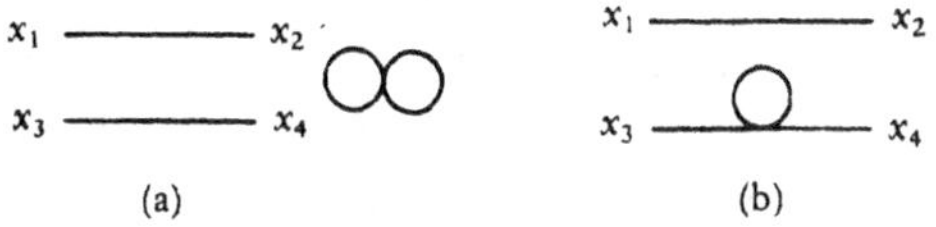

FIG. 15

expressions

$$\Delta_F(x_1-x_2)\,\Delta_F(x_3-x_4)\lambda\int dx\,\Delta_F(0)\,\Delta_F(0)$$

and

$$\lambda\,\Delta_F(x_1-x_2)\int dx\,\Delta_F(x_3-x)\,\Delta_F(x-x_4)\,\Delta_F(0)$$

respectively. This method is essentially that of coordinate space Feynman diagrams and is the most convenient way of displaying the results of differentiations with respect to J. Now of all the diagrams in Figs 13 and 14 only one is connected, namely the last one appearing in Fig. 14. It stands for the expression

$$-\lambda\int dx\,\Delta_F(x_1-x)\,\Delta_F(x_2-x)\,\Delta_F(x_3-x)\,\Delta_F(x_4-x). \quad (1.183)$$

This expression is the contribution to $\langle 0|T(\phi(x_1)\phi(x_2)\phi(x_3)\phi(x_4))|0\rangle_C$; thus the contribution $\langle 0|T(\phi(x_1)\phi(x_2)\phi(x_3)\phi(x_4))|0\rangle_C$ is, using 1.175,176,

$$-\lambda\int dx\,\delta(x_1-x)\,\delta(x_2-x)\,\delta(x_3-x)\,\delta(x_4-x)$$
$$=-\lambda\,\delta(x_2-x_1)\,\delta(x_3-x_1)\,\delta(x_4-x_1). \quad (1.184)$$

Finally, $T(p_1\ldots p_4)$ is found to be simply $-i\lambda$ as required. In this way we can verify that functional integration yields the entire perturbation series. Before leaving this calculation we wish to make a few important points. The first is about the diagrams with no external lines; these are usually called vacuum graphs. It is clear that since the denominator of $Z[J]$ does not depend on J its expansion in powers of λ consists entirely of vacuum graphs. Vacuum graphs also occur as we have seen as contributions to the disconnected parts of Green's functions when the numerator is expanded in powers of λ. Now in our example the vacuum graphs cancelled when the expansions of the numerator and denominator were combined. This left only graphs, some disconnected and some connected, with external lines. This cancellation of vacuum graphs between the denominator and numerator is a general result and occurs for all Green's functions to all orders in λ. This is an advantage of the functional method. When calculating Feynman diagrams using the usual methods, vacuum diagrams occur in the expansion of Wick products. They have to be simply discarded as representing an unmeasurable phase for S matrix elements. Here that phase is already divided out. The second point concerns the diagrams which contain factors of $\Delta_F(0)$, e.g. Fig. 15(b). Figure 15(a) also contains factors of $\Delta_F(0)$ but it has cancelled away as we have just seen. Such diagrams are infinite because $\Delta_F(0)$ is infinite. Although we shall encounter many infinities in renormalizing Feynman diagrams, '$\Delta_F(0)$ infinities' may be removed from the beginning. Their removal is accomplished by replacing the interaction term $(\lambda \phi^4/4!)(x)$ by the normal product term denoted by $\lambda/4!:\phi^4(x):$. This is a procedure done when calculating Feynman diagrams by more conventional methods as well. Before reminding the reader of the definition of the normal product we shall summarize its effect on the diagrams. The consequence of normal ordering is to remove all diagrams which contain vertices from which a line both begins and ends. This clearly just means diagrams which give rise to the factor $\Delta_F(0)$. The normal product is defined first for two operators multiplied together then for three etc. The normal product of two operators is defined by

$$:\phi(x)\phi(y): = T(\phi(x)\phi(y)) - \langle 0|T(\phi(x)\phi(y))|0\rangle \quad (1.185)$$

for three by

$$:\phi(x)\phi(y)\phi(z): = T(\phi(x)\phi(y)\phi(z)) - \phi(x)\langle 0|T(\phi(y)\phi(z))|0\rangle$$
$$- \phi(y)\langle 0|T(\phi(x)\phi(z))|0\rangle$$
$$- \phi(z)\langle 0|T(\phi(x)\phi(y))|0\rangle \quad (1.186)$$

and for four by†

$$:\phi(x)\phi(y)\phi(z)\phi(t): = T(\phi(x)\phi(y)\phi(z)\phi(t))$$
$$- \langle 0|T(\phi(x)\phi(y))|0\rangle:\phi(z)\phi(t):$$
$$- \langle 0|T(\phi(x)\phi(z))|0\rangle:\phi(y)\phi(t): - \langle 0|T(\phi(x)\phi(t))|0\rangle:\phi(y)\phi(z):$$
$$- \langle 0|T(\phi(z)\phi(t))|0\rangle:\phi(x)\phi(y): - \langle 0|T(\phi(y)\phi(t))|0\rangle:\phi(x)\phi(z):$$
$$- \langle 0|T(\phi(y)\phi(z))|0\rangle:\phi(x)\phi(t):$$
$$- \langle 0|T(\phi(x)\phi(y))|0\rangle\langle 0|T(\phi(z)\phi(t))|0\rangle$$
$$- \langle 0|T(\phi(x)\phi(z))|0\rangle\langle 0|T(\phi(y)\phi(t))|0\rangle$$
$$- \langle 0|T(\phi(x)\phi(t))|0\rangle\langle 0|T(\phi(z)\phi(y))|0\rangle. \quad (1.187)$$

† The vacuum expectation value in these formulae stands for that given in 1.172 for the case $n = 2$.

We require $:\phi^4(x):$; this is accomplished by taking the limit of 1.187 as x, y, z, t all tend to the same value. Incidentally, note that the time ordering symbol may be removed when $x = y = z = t$. It is now simple to see that the terms with $\Delta_F(0)$ are being subtracted out of the product $\phi^4(x)$. It may now be checked that this normal ordering procedure does get rid of all diagrams like Fig. 15(b). Before leaving the subject of normal ordering we note that the vacuum graph occurring in Fig. 15(a) is also removed by normal ordering. However, this does not mean that the above argument about cancellation of vacuum graphs is unnecessary. This is because in higher orders there will occur more complicated graphs. For example to order λ^2 there is the

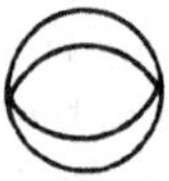

FIG. 16

graph of Fig. 16. This graph represents the expression

$$\lambda^2 \int dx\, dy\, \Delta_F^4(x-y), \qquad (1.188)$$

an expression which is not removed by normal ordering, and only disappears in the cancellation between numerator and denominator described above. The generalization of the normal product formula to give the normal product of n operators should be clear from 1.185, 1.186, 1.187. However, we give the general formula here. It has a compact form and is very convenient to use. It is

$$:\phi(x_1)\ldots\phi(x_n): = T(\phi(x_1)\ldots\phi(x_n))$$

$$- \left[\exp\left\{\frac{1}{2}\int dx\, dy\, \langle 0|T(\phi(x)\phi(x))|0\rangle \frac{\delta}{\delta\phi(x)}\frac{\delta}{\delta\phi(y)}\right\} - 1\right]$$

$$:\phi(x_1)\ldots\phi(x_n):$$

$$= T(\phi(x_1)\ldots\phi(x_n))$$

$$- \left[\exp\left\{\frac{i}{2}\int dx\, dy\, \Delta_F(x-y)\frac{\delta}{\delta\phi(x)}\frac{\delta}{\delta\phi(y)}\right\} - 1\right]$$

$$.:\phi(x_1)\ldots\phi(x_n):. \qquad (1.189)$$

Functional Integrals for Fermions

Just as the theory of functional differentiation changes slightly as one changes from Bose to Fermi fields so does the theory of functional integration. Functional differentiation for Fermi fields means, as we have seen, differentiation on an infinite-dimensional Grassmann algebra. Functional integration for Fermi fields means defining integration on an infinite-dimensional Grassmann algebra. First of all we must define integration over a finite-dimensional Grassmann algebra. Let G be a Grassmann algebra of dimension 2^n, i.e. satisfying 1.92, 1.93. It is up to us to *define* an integral on this algebra which makes sense and corresponds as closely as possible to our intuitive notion of an integral. We shall simply present the definitions and show how they give rise to a satisfactory integration theory. A volume element is written $dC_1 \ldots dC_i$, $1 \le i \le n$, where C_i and dC_i must satisfy

$$\{dC_i, dC_j\} = 0 \qquad \{dC_i, C_j\} = 0. \tag{1.190}$$

The integral is then defined by defining

$$\int dC_i = 0 \qquad \int C_i \, dC_i = 1. \tag{1.191}$$

These are peculiar looking formulae but in view of the anti-commuting nature of G some departure from conventional looking formulae should be expected. Equations 1.190, 1.191 are sufficient to allow us to integrate any elements of the algebra G. For example, let $g(C)$ be the element of G given in 1.94. Then 1.190, 1.191 tell us immediately that

$$\int g(C) \, dC_1 \ldots dC_i = (-1)^i \, i! \, G^i_{1,2\ldots i} \qquad 1 \le i \le n \tag{1.192}$$

where $G^i_{1,2\ldots i}$ is assumed to be antisymmetric under interchange of $1, 2 \ldots i$. Take as a further illustration $i = 3$. Then the first formula of 1.190 tells us that

$$\int g(C) \, dC_1 \, dC_2 \, dC_3 = G^3_{i_1 i_2 i_3} \int C_{i_1} C_{i_2} C_{i_3} \, dC_1 \, dC_2 \, dC_3, \tag{1.193}$$

but

$$G^3_{i_1 i_2 i_3} C_{i_1} C_{i_2} C_{i_3} = G^3_{123} C_1 C_2 C_3 + G^3_{132} C_1 C_3 C_2 + G^3_{231} C_2 C_3 C_1$$
$$+ G^3_{213} C_2 C_1 C_3 + G^3_{312} C_3 C_1 C_2 + G^3_{321} C_3 C_2 C_1$$
$$= 6 G^3_{123} C_1 C_2 C_3, \tag{1.194}$$

Hence

$$\int g(C)\, dC_1\, dC_2\, dC_3 = 3!\, G_{123}^3 \int C_1 C_2 C_3\, dC_1\, dC_2\, dC_3$$

$$= -3!\, G_{123}^3 \int C_3 C_2 C_1\, dC_1\, dC_2\, dC_3$$

$$= -3!\, G_{123}^3. \tag{1.195}$$

These examples make the handling of integrals quite straightforward. A note of caution though on change of variables during integration. The usual formula for change of variables using a Jacobian is invalid. We shall now explain this. Let $f(x_1 \ldots x_m)$ be a function of m real variables. Define the integral I of f by

$$I = \int f(\mathbf{x})\, dx_1 \ldots dx_m. \tag{1.196}$$

If M is an $m \times m'$ matrix we can also write, ($\det M \neq 0$),

$$\mathbf{x} = M\mathbf{y} \qquad I = \det M \int f(M\mathbf{y})\, dy_1 \ldots dy_m. \tag{1.197}$$

However, let

$$I = \int g(C)\, dC_1 \ldots dC_m, \tag{1.198}$$

define the linear change of variable in the Grassmann algebra by

$$\mathbf{C} = M\mathbf{e} \tag{1.199}$$

then we find that

$$I = (\det M)^{-1} \int g(M\mathbf{e})\, de_1 \ldots de_m. \tag{1.200}$$

The 'Jacobian' is the inverse of what we expected. The reason for this is easy to find. It is that

$$C_1 \ldots C_m = (\det M) e_1 \ldots e_m \tag{1.201}$$

but

$$dC_1 \ldots dC_m = (\det M)^{-1} de_1 \ldots de_m \tag{1.202}$$

so that $\int C_i\, dC_i = 1$ may be maintained.

For use in field theories with Fermions we require an infinite-dimensional algebra G. We recall that in the infinite-dimensional case we wrote the anti-commuting variables as $C(x)$ and expressed a general element of G by the functional power series

$$P(C(x)) = C_0 + \int C_1(x_1) C(x_1)\, d(x_1) + \cdots$$
$$+ \int C_n(x_1 \ldots x_n) C(x_1) \ldots C(x_n)\, dx_1 \ldots dx_n + \cdots. \quad (1.203)$$

The integration theory will be constructed by defining

$$dC(x) = C(x)\, dx. \quad (1.204)$$

With this definition it follows that

$$\{dC(x), dC(x)\} = dx\, dx'\{C(x), C(x')\}$$
$$= 0 \quad (1.205)$$

and

$$\{dC(x'), C(x)\} = dx'\{C(x'), C(x)\}$$
$$= 0. \quad (1.206)$$

We now only need to normalize the variables $C(x)$ by restricting them so that

$$\int C(x)\, dC(x) = 1 \qquad \int dC(x) = 0. \quad (1.207)$$

The infinite-dimensional integral is obtained by the usual limiting procedure of starting with a finite-dimensional integral (in this case also a finite-dimensional Grassmann algebra), and taking the limit as n goes to infinity. If $F(C)$ is a functional to be integrated over G we shall write the integral as

$$\int F(C) \prod dC(x). \quad (1.208)$$

When working with quantum theories with Fermions we shall use the more conventional notation $\mathscr{D}[C(x)]$ instead of $\prod dC(x)$. As might be expected the functional integral that is most important for Fermi fields is the integral of the exponential of iS where S is the action. The action for free Fermions is given by

$$S(\psi, \bar{\psi}) = \int dx\, \bar{\psi}(x)(i\not{\partial} - m)\psi. \quad (1.209)$$

In the functional integral ψ and $\bar{\psi}$ are independent integration variables. This means that the analogue of $F[J]$ is the functional $F[\zeta, \bar{\zeta}]$. Where

$$F[\zeta, \bar{\zeta}] = \int \mathcal{D}[\psi]\mathcal{D}[\bar{\psi}] \exp\left[i\int dx\, \{\bar{\psi}(x)(i\not{\partial} - m)\psi + \bar{\zeta}\psi + \bar{\psi}\zeta\}\right], \quad (1.210)$$

and $\zeta(x)$ and $\bar{\zeta}(x)$ are the sources of $\bar{\psi}(x)$ and $\psi(x)$ respectively and are elements of the Grassmann algebra. Using exactly the same techniques as those used to evaluate $Z[J]$ it may be shown that

$$F[\zeta, \bar{\zeta}] = \exp\left[\int dx\, dy\, (-i)\bar{\zeta}(x) S_F(x-y) \zeta(x)\right] \cdot D^{-1} \int \mathcal{D}[\psi]\mathcal{D}[\bar{\psi}] \exp\left[\int i\bar{\psi}\psi\, dx\right] \quad (1.211)$$

where

$$S_F(x) = (i\not{\partial} + m)\Delta_F(x).$$

Note that in all respects this is a very similar formula to that for $F[J]$ in 1.166. An important difference is that the Fredholm determinant D occurs as D^{-1} in 1.211 rather than D as it does in 1.166. This is because of the change of variable formula given in 1.200. Just as we normalized $F[J]$ to $Z[J]$ so that $Z[0] = 1$ we define $Z[\zeta, \bar{\zeta}]$ by

$$Z(\zeta, \bar{\zeta}) = \frac{\int \mathcal{D}[\psi]\mathcal{D}[\bar{\psi}] \exp\left[i\int dx\, \{\bar{\psi}(x)(i\not{\partial} - m)\psi + \bar{\zeta}\psi + \bar{\psi}\zeta\}\right]}{\int \mathcal{D}[\psi]\mathcal{D}[\bar{\psi}] \exp\left[i\int dx\, \{\bar{\psi}(i\not{\partial} - m)\psi\}\right]}. \quad (1.212)$$

Now $Z[0, 0] = 1$ (note that the Fredholm determinant D^{-1} cancels in the ratio). Having already shown in some detail how the functional integral technique gives rise to the perturbation series of Feynman diagrams we shall not repeat it here for QED. We shall state the result for the general Green's function. It is that

$$\langle 0|T(A_{\mu_1}(x_1)\ldots A_{\mu_n}(x_n)\bar{\psi}(y_1)\ldots\bar{\psi}(y_m)\psi(z_1)\ldots\psi(z_m))|0\rangle$$

$$= (-i)^{n+2m} \frac{\delta^{n+2m} Z[\zeta, \bar{\zeta}, J]}{\delta J_{\mu_1}(x_1)\ldots \delta J_{\mu_n}(x_n) \delta\zeta(y_1)\ldots \delta\zeta(y_m) \delta\bar{\zeta}(z_1)\ldots \delta\bar{\zeta}(z_m)} \quad (1.213)$$

with $J_\mu = \zeta = \bar\zeta = 0$ where

$$Z[\zeta,\bar\zeta,J] = \Bigg[\int \mathcal{D}[\psi]\mathcal{D}[\bar\psi]\mathcal{D}[A]\exp\Big[i\int dx\,\{-\tfrac{1}{4}F_{\mu\nu}F^{\mu\nu}$$
$$+ \bar\psi(i\slashed{\partial}-m)\psi - \tfrac{1}{2}(\partial_\lambda A^\lambda)^2$$
$$- e:\bar\psi\gamma_\mu A^\mu\psi: + J_\mu A^\mu + \bar\zeta\psi + \bar\psi\zeta\}\Big]\Bigg]\Bigg[\int \mathcal{D}[\psi]\mathcal{D}[\bar\psi]\mathcal{D}[A]$$
$$\exp\Big[i\int dx\,\{-\tfrac{1}{4}F_{\mu\nu}F^{\mu\nu} + \bar\psi(i\slashed{\partial}-m)\psi - \tfrac{1}{2}(\partial_\lambda A^\lambda)^2$$
$$- e:\bar\psi\gamma_\mu A^\mu\psi:\}\Big]\Bigg]^{-1} \quad (1.214)$$

In 1.214 $J_\mu(x)$ is a C number source for the photon field $A_\mu(x)$. When $e=0$, the functional integrals over ψ, $\bar\psi$ and A_μ can all be reduced to integrals over quadratic forms with the result

$$Z[\zeta,\bar\zeta,J] = \exp\Big[-i\int dx\,dy\,\bar\zeta(x)S_F(x-y)\zeta(y)\Big]$$
$$\cdot \exp\Big[-\frac{i}{2}\int dx\,dy\,J_\mu(x)D_F^{\mu\nu}(x-y)J_\nu(y)\Big] \quad (1.215)$$

for $e = 0$. The function $D_F^{\mu\nu}(x)$ is the photon Green's function defined by

$$D_F^{\mu\nu}(x) = g^{\mu\nu}D_F(x) \quad (1.216)$$

where $D_F(x)$ is the Green's function for the zero mass Klein–Gordon equation

$$\Box D_F(x-y) = -\delta(x-y). \quad (1.217)$$

For $e > 0$ we remind the reader that we must, as we have indeed done, normal-order the coefficient of e in the action S. The Feynman diagrams of QED are then automatically generated by the functional integral technique. The normal ordering of products of Fermion fields proceeds in close analogy with that for Bose fields. The general formula for the Fermi case which replaces 1.189 is

$$:\bar\psi(x_1)\ldots\bar\psi(x_n)\psi(y_1)\ldots\psi(y_n): = T(\bar\psi(x_1)\ldots\bar\psi(x_n)\psi(y_1)\ldots\psi(y_n))$$
$$-\Big[\exp\Big\{i\int dx\,dy\,S_F(x-y)\frac{\delta}{\delta\bar\psi(x)}\frac{\delta}{\delta\psi(y)}\Big\} - 1\Big]$$
$$.:\bar\psi(x_1)\ldots\bar\psi(x_n)\psi(y_1)\ldots\psi(y_n):. \quad (1.218)$$

Generating Functions

The functionals $Z[J]$ and $Z[\zeta, \bar{\zeta}, J_\mu]$ are usually called generating functions since they can be used to 'generate' the series of Feynman graphs. As we have seen, they generate not only connected graphs but also disconnected graphs (i.e. graphs that do not hang together but have two or more separate pieces). It is possible to find a functional $W[J]$ simply related to $Z[J]$ that generates only the connected graphs or connected Green's functions. $W[J]$ is related to $Z[J]$ by the equation

$$W[J] = -i \ln Z[J] \qquad (1.219)$$

or

$$Z[J] = \exp[iW[J]]. \qquad (1.220)$$

The swiftest method of verifying that $W[J]$ generates the connected Green's functions is by functional differentiation. Evidently

$$\frac{\delta^2 W[J]}{\delta J(x_1) \delta J(x_2)} = \frac{i}{Z^2} \frac{\delta Z}{\delta J(x_2)} \frac{\delta Z}{\delta J(x_1)} - \frac{i}{Z} \frac{\delta^2 Z}{\delta J(x_1) \delta J(x_2)}. \qquad (1.221)$$

We assume $\langle \phi(x) \rangle = 0$ so that if we set $J = 0$ then $\delta Z/\delta J(x)$ is zero and $Z[0]$ is unity so we have

$$\frac{\delta^2 W[0]}{\delta J(x_1) \delta J(x_2)} = -i \frac{\delta^2 Z}{\delta J(x_1) \delta J(x_2)} = i \langle 0 | T(\phi(x_1)\phi(x_2)) | 0 \rangle. \qquad (1.222)$$

Thus $-i(\delta^2 W[0]/\delta J(x_1) \delta J(x_2))$ is the propagator which in any case has no disconnected part. Proceeding to the next non-zero Green's function we find by differentiating 1.221 and setting $J = 0$ that

$$\frac{\delta^4 W[0]}{\delta J(x_1) \delta J(x_2) \delta J(x_3) \delta J(x_4)} = i \frac{\delta^2 Z[0]}{\delta J(x_1) \delta J(x_2)} \frac{\delta^2 Z[0]}{\delta J(x_3) \delta J(x_4)}$$
$$+ i \frac{\delta^2 Z[0]}{\delta J(x_1) \delta J(x_3)} \frac{\delta^2 Z[0]}{\delta J(x_2) \delta J(x_4)}$$
$$+ i \frac{\delta^2 Z[0]}{\delta J(x_1) \delta J(x_4)} \frac{\delta^2 Z[0]}{\delta J(x_2) \delta J(x_3)} - i \frac{\delta^4 Z[0]}{\delta J(x_1) \delta J(x_2) \delta J(x_3) \delta J(x_4)}. \qquad (1.223)$$

But the RHS of 1.223 is simply

$$i[\langle 0|T(\phi(x_1)\phi(x_2))|0\rangle\langle 0|T(\phi(x_3)\phi(x_4))|0\rangle$$
$$+\langle 0|T(\phi(x_1)\phi(x_3))|0\rangle\langle 0|T(\phi(x_2)\phi(x_4))|0\rangle$$
$$+\langle 0|T(\phi(x_1)\phi(x_4))|0\rangle\langle 0|T(\phi(x_2)\phi(x_3))|0\rangle$$
$$-\langle 0|T(\phi(x_1)\phi(x_2)\phi(x_3)\phi(x_4))|0\rangle]. \quad (1.224)$$

Now if we refer to our definition of the connected Green's function in 1.174 we see that 1.224 is just the expression $-i\langle 0|T(\phi(x_1)\phi(x_2)\phi(x_3)\phi(x_4))|0\rangle_C$. In terms of W the connected Green's function is simply

$$i\frac{\delta^4 W[0]}{\delta J(x_1)\,\delta J(x_2)\,\delta J(x_3)\,\delta J(x_4)}. \quad (1.225)$$

Induction can be used to make this argument general. The formula for the connected Green's functions that results is

$$\langle 0|T(\phi(x_1)\ldots\phi(x_n))|0\rangle_C = (-i)^{n-1}\frac{\delta^n W[0]}{\delta J(x_1)\ldots\delta J(x_n)}. \quad (1.226)$$

For free fields 1.167 tells us that $W[J] = -\tfrac{1}{2}\int dx\,dy\,J(x)\Delta_F(x-y)J(y)$. This means that for $n > 2$ the connected Green's functions are all zero corresponding to the absence of scattering due to the absence of interaction.

Another useful generating function may be derived from $W[J]$. It is written $\Gamma[\phi]$ and is called the generator of one particle irreducible amputated graphs of Green's functions. The word amputation has the same meaning as it did in 1.175. The amputated Green's functions have no propagators attached to their external legs. The phrase one particle irreducible is more easily understood in graphical language. A Feynman graph is said to be one-particle reducible if by cutting any one of the internal lines the graph is not made up of two disconnected pieces. We show this schematically in Fig. 17. The bubbles in Fig. 17

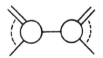

FIG. 17

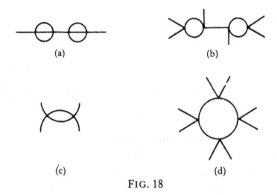

FIG. 18

stand for arbitrarily complex Feynman graphs. Figures 18(a) and (b) are two simple examples of one particle reducible graphs, while Figs 18(c) and (d) are one particle irreducible. Being able to generate one-particle irreducible graphs is useful as in graphical analyses it is frequently necessary to categorize graphs by this property. One-particle irreducible (we shall not bother to include the word amputated as well) Green's functions† also appear in certain exact equations of field theory which we shall discuss in the next section. The definition of $\Gamma[\phi]$ is given by performing a functional Legendre transformation on $W[J]$

$$W[J] = \Gamma[\phi] + \int dx\, J(x)\phi(x). \tag{1.227}$$

The variables $\phi(x)$ and $J(x)$ satisfy the equations

$$\frac{\delta W[J]}{\delta J(x)} = \phi(x), \qquad \frac{\delta \Gamma[\phi]}{\delta \phi(x)} = -J(x). \tag{1.228}$$

From this we deduce the two equations

$$\frac{\delta^2 W[J]}{\delta J(x)\, \delta J(y)} = \frac{\delta \phi(x)}{\delta J(y)}, \qquad \frac{\delta^2 \Gamma[\phi]}{\delta \phi(x)\, \delta \phi(y)} = -\frac{\delta J(x)}{\delta \phi(y)}, \tag{1.229}$$

then we discover that

$$\int d\xi\, \frac{\delta^2 W[J]}{\delta J(x)\, \delta J(\xi)}\, \frac{\delta^2 \Gamma[\phi]}{\delta \phi(\xi)\, \delta \phi(y)} = -\int d\xi\, \frac{\delta \phi(x)}{\delta J(\xi)}\, \frac{\delta J(\xi)}{\delta \phi(y)}$$

$$= -\frac{\delta \phi(x)}{\delta \phi(y)} = -\delta(x - y). \tag{1.230}$$

† We shall also use the term proper vertices to describe such Green's functions.

In other words $-\delta^2\Gamma[\phi]/\delta\phi(\xi)\,\delta\phi(y)$ is the inverse of $\delta^2 W[J]/\delta J(x)\,\delta J(\xi)$. Hence if we set $J=0$ then $-i(\delta^2 W[0]/\delta J(x)\,\delta J(y))$ is the propagator and therefore $-i(\delta^2\Gamma[0]/\delta\phi(x)\,\delta\phi(y))$ is the inverse propagator. The next property of $\Gamma[\phi]$ is obtained by differentiating 1.230 with respect to $J(z)$. We obtain

$$\int d\xi \frac{\delta^3 W[J]}{\delta J(x_1)\,\delta J(\xi)\,\delta J(z)} \frac{\delta^2\Gamma[\phi]}{\delta\phi(\xi)\,\delta\phi(y)}$$
$$+ \int d\xi\, d\xi_1 \frac{\delta^2 W[J]}{\delta J(x)\,\delta J(\xi)} \frac{\delta^3\Gamma[\phi]}{\delta\phi(\xi)\,\delta\phi(y)\,\delta\phi(\xi_1)} \frac{\delta\phi(\xi_1)}{\delta J(z)} = 0. \quad (1.231)$$

But in view of 1.229 and 1.230 this equation implies immediately that

$$\frac{\delta^3\Gamma[\phi]}{\delta\phi(x)\,\delta\phi(y)\,\delta\phi(z)} = -\int dx'\,dy'\,dz' \frac{\delta^2\Gamma[\phi]}{\delta\phi(x)\,\delta\phi(x')} \frac{\delta^2\Gamma[\phi]}{\delta\phi(y)\,\delta\phi(y')}$$
$$\cdot \frac{\delta^2\Gamma[\phi]}{\delta\phi(z)\,\delta\phi(z')} \frac{\delta^3 W[J]}{\delta J(x')\,\delta J(y')\,\delta J(z')}. \quad (1.232)$$

But since W generates connected Green's functions and $-i(\delta^2\Gamma/\delta\phi(x)\,\delta\phi(y))$ is the inverse propagator then $i(\delta^3\Gamma/\delta\phi(x)\,\delta\phi(y)\,\delta\phi(z))$ is simply the amputated connected three-point function. We can represent this diagrammatically by Fig. 19.

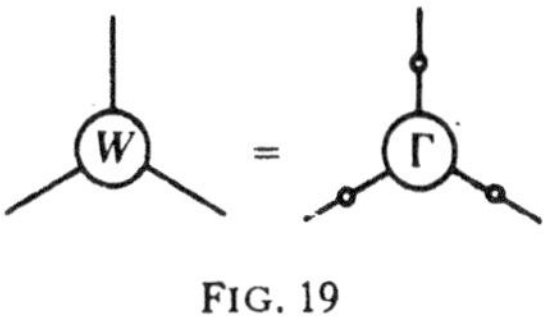

FIG. 19

The straight line with the circle in the middle represents the propagator attached to the external lines of the Green's function Γ. For the $\lambda\phi^4/4!$ field theory which we have been working with there is no connected Green's function with three external legs. Hence if we set $J=0$ in 1.232 the equation becomes $0=0$. However, if we were to consider the field theory with interaction theory term $-\lambda\phi^3/3!$ rather than $-\lambda\phi^4/4!$ then this is no longer so. For $\lambda\phi^3/3!$ Green's functions are non-zero whether the number of external legs is odd or even. The Feynman graphs of $\lambda\phi^3/3!$ theory differ from those of $\lambda\phi^4/4!$ theory only in the structure of the vertex. The vertex in $\lambda\phi^3/3!$ theory consists of three rather than four lines coming to a point. In Fig. 20 we show the two lowest order graphs contributing to the three point function in $\lambda\phi^3/3!$ theory.

FIG. 20

To proceed further we could differentiate 1.232 with respect to $\phi(\hat{w})$. It is, however, simpler to use the property of inverse contained in 1.230 to write 1.232 as an equation for $\delta^3 W/\delta J(x')\,\delta J(y')\,\delta J(z')$ and then differentiate this with respect to $J(w)$. The resultant expression will also contain a third order derivative of W and this must be eliminated using 1.232. The most compact way of representing the final equation is by diagrams. We display the diagrammatic equation in Fig. 21. For example the lowest order graphs in $\lambda\phi^3/3!$

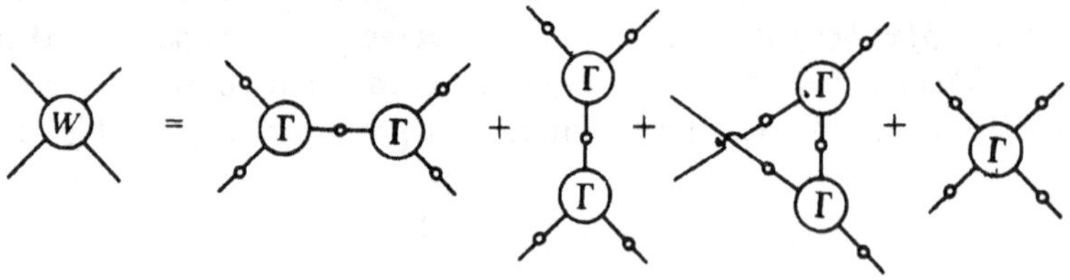

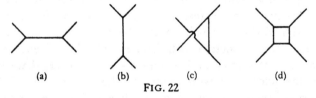

FIG. 21

theory would give rise to the one-particle reducible diagrams of Fig. 22(a), (b) and (c), and the one-particle irreducible diagram of Fig. 22(d).

FIG. 22

As is usual for the properties of generating functions the proof that $i(\delta^n\Gamma[0]/\delta\phi(x_1)\ldots\delta\phi(x_n))$ is the amputated one-particle irreducible Green's function can be done by induction. This completes our list of generating functions. We shall just state that for QED the connected Green's functions are given by

$$\frac{(-i)^{n+2m-1}\delta^{n+2m}W[0]}{\delta J_{\mu_1}(x_1)\ldots\delta J_{\mu_n}(x_n)\,\delta\bar{\zeta}(y_1)\ldots\delta\bar{\zeta}(y_n)\,\delta\zeta(z_1)\ldots\delta\zeta(z_m)}$$

where

$$W[\zeta,\bar{\zeta},J] = -i \ln Z[\zeta,\bar{\zeta},J]. \qquad (1.233)$$

The amputated one-particle irreducible Green's functions are given by

$$i\frac{\delta^{n+2m}\Gamma[0]}{\delta A_{\mu_1}(x_1)\ldots\delta A_{\mu_n}(x_n)\,\delta\bar{\psi}(y_1)\ldots\delta\bar{\psi}(y_m)\,\delta\psi(y_1)\ldots\delta\psi(y_m)}$$

where

$$\Gamma[\bar{\psi},\psi,A] = W[\zeta,\bar{\zeta},J] - \int dx\,\bar{\psi}\zeta - \int dx\,\bar{\zeta}\psi - \int dx\,J_\mu A^\mu. \qquad (1.234)$$

The Schwinger–Dyson Equations

The Schwinger–Dyson equations are certain integral equations obeyed by the Green's functions of quantum field theories. They are of interest because they are obtained without any approximations procedure and hence are exact. This means that they can in principle be used to study the non-perturbative modes as well as the perturbative mode of a particular field theory. It is only fair to say though that very few useful non-perturbative results have been obtained from their study. We derive first the equations for $\lambda\phi^4/4!$ theory. The integral equation which is our goal is derived by first obtaining a functional differential equation, and then integrating it. The derivation of the functional differential equation is accomplished by the technique of functional Fourier transformation. The functional Fourier transform of a functional $\hat{A}[\phi]$ is written $A[J]$ and defined by

$$A[J] = \int \mathcal{D}[\phi]\hat{A}[\phi]\exp\left[i\int dx\,J(x)\phi(x)\right]. \qquad (1.235)$$

With this definition we turn to 1.173 for the definition of $Z[J]$; we see straight away that $Z[J]$ is the functional Fourier transform of the functional $\hat{Z}[\phi]$ where

$$\hat{Z}[\phi] \doteq \exp\left[i\int dx\left\{\frac{(\partial_\mu\phi)^2}{2} - \frac{m^2\phi^2}{2} - \frac{:\lambda\phi^4:}{4!}\right\}\right]\Big/D \qquad (1.236)$$

(D stands for the denominator of 1.173). By functional differentiation we deduce that

$$i\frac{\delta \hat{Z}[\phi]}{\delta \phi(x_1)} = (\Box + m^2)\phi(x_1)\hat{Z}[\phi] + \frac{\lambda}{3!}:\phi^3(x_1):\hat{Z}[\phi]. \quad (1.237)$$

It is a simple matter to invert the Fourier transform and obtain an equation for $Z[J]$. Using techniques familiar from the Fourier theory for ordinary functions we find that $Z[J]$ satisfies the equation†

$$(\Box + m^2)\frac{\delta Z[J]}{\delta J(x_1)} = -iJ(x_1)Z[J] + \frac{\lambda}{3!}\frac{:\delta^3 Z:}{\delta J(x_1)\,\delta J(x_1)\,\delta J(x_1)}Z[J]. \quad (1.238)$$

This equation is further differentiated with respect to $J(x_2)$ to yield, after setting $J = 0$, the result

$$(\Box + m^2)\langle 0|T(\phi(x_1)\phi(x_2))|0\rangle = i\,\delta(x_1 - x_2) - \frac{\lambda}{3!}\langle 0|T(:\phi^3(x_1):\phi(x_2))|0\rangle. \quad (1.239)$$

This equation may be solved using the Green's function Δ_F. The solution takes the form

$$\langle 0|T(\phi(x_1)\phi(x_2))|0\rangle = \int d\xi\,\Delta_F(x_1 - \xi)i\,\delta(\xi - x_2)$$

$$- \frac{\lambda}{3!}\int d\xi\,\Delta_F(x_1 - \xi)\langle 0|T(:\phi^3(\xi):\phi(x_2))|0\rangle$$

$$= i\,\Delta_F(x_1 - x_2) + \frac{\lambda}{3!}\int d\xi\,\Delta_F(x_1 - \xi)\langle 0|T(:\phi^3(\xi):\phi(x_2))|0\rangle. \quad (1.240)$$

It is possible to proceed more quickly if we use a diagrammatic representation of the solution. In Fig. 23 the LHS is the propagator; the straight lines with the labels x_1, x_2 and x_1, ξ at each end stand for the free propagator $\Delta_F(x_1 - x_2)$ and $\Delta_F(x_1 - \xi)$; the bubble with four

† The symbol $:\delta^3 Z:/\delta J(x_1)\,\delta J(x_1)\,\delta J(x_1)$ stands for
$$\frac{\int \mathscr{D}[\phi]:\phi^3(x_1):\exp[iS(\phi) + i\int J\phi]}{\int \mathscr{D}[\phi]\exp[iS(\phi)]}$$
where $S(\phi)$ is the action.

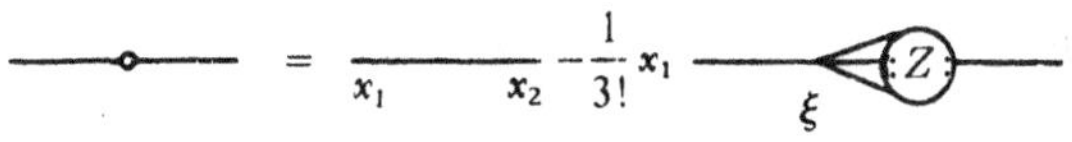

$$\text{———o———} = \frac{1}{x_1 \quad x_2} - \frac{1}{3!} x_1 \text{———}\triangleleft{:Z:}\text{———}$$
$$\xi$$

FIG. 23

external lines and the symbol $:Z:$ inside stands for the contribution of the four-point function disconnected and connected. The normal ordering reminds us that the term $\lambda\phi^4/4!$ in the action was normal ordered. This is important; to see this consider the case when normal ordering is not done. The decomposition of the four-point function into its disconnected and connected parts gives

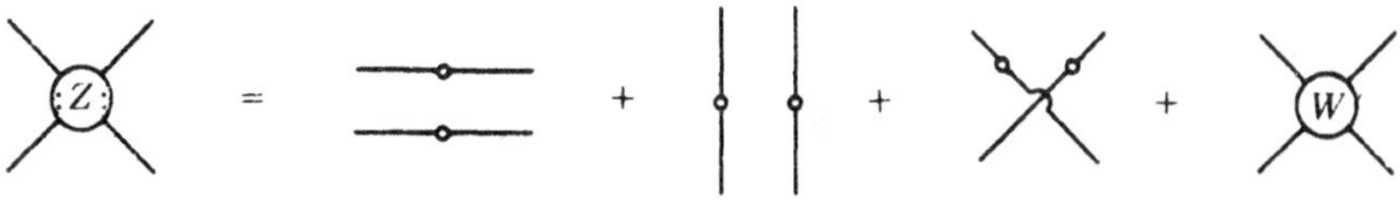

FIG. 24

where W indicates the connected part. With this decomposition the second diagram in Fig. 23 becomes

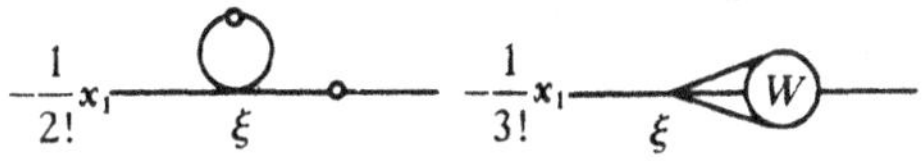

$$-\frac{1}{2!}x_1 \text{——}\bigcirc\text{——} -\frac{1}{3!}x_1 \text{——}\triangleleft{W}\text{——}$$

FIG 25

The first term in Fig. 25 contains $\Delta_F(0)$. Fortunately since we have carried out normal ordering this term is absent and so we may write

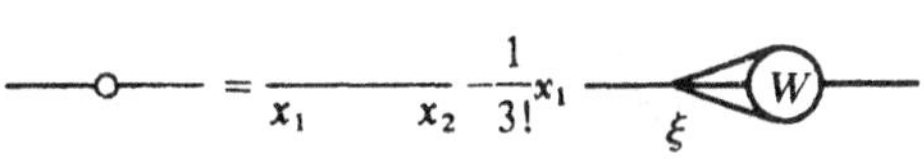

$$\text{———o———} = \frac{1}{x_1 \quad x_2} - \frac{1}{3!}x_1 \text{———}\triangleleft{W}\text{———}$$
$$\xi$$

FIG. 26

The final result will take on a simpler form in momentum space since the convolutions become products. We therefore use Fourier transform with respect to x_1 and x_2 and factor off the resultant energy momentum conserving delta function that will then multiply the LHS and RHS. We write the result as

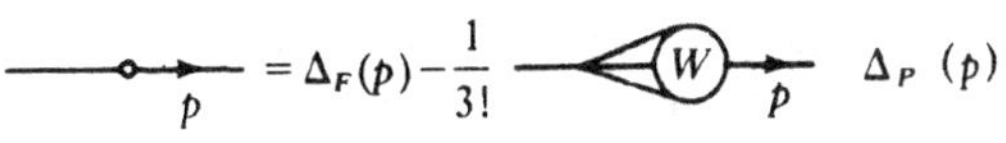

$$\text{——o——}_p = \Delta_F(p) - \frac{1}{3!} \text{——}\triangleleft{W}\text{——}_p \; \Delta_F(p)$$

FIG. 27

It should now be noted that the propagator from x_1 to ξ has been removed from the external line and written explicitly as $\Delta_F(p)$. The next step is to use the relation between W and Γ. In $\lambda \delta^4/4!$ theory the first three terms in Fig. 21 are zero because they involve Green's functions with odd numbers of external legs. The relation between W and Γ is therefore

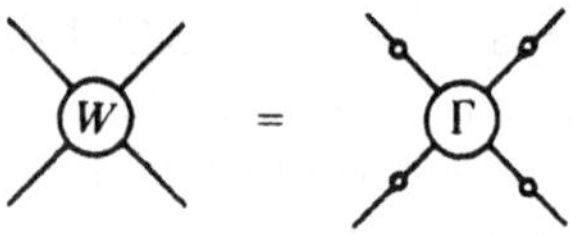

FIG. 28

The equation in Fig. 27 now takes on the form

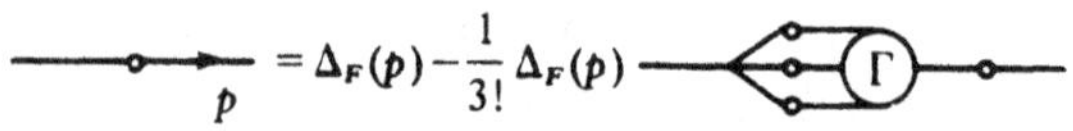

FIG. 29

Dividing across by ─o─ and $\Delta_F(p)$ we obtain the equation for the inverse propagator

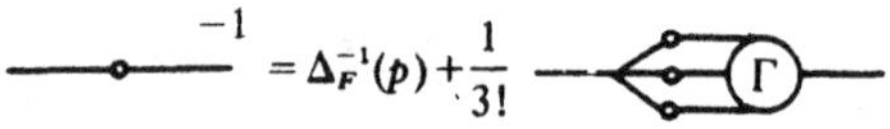

FIG. 30

This is known as the Schwinger–Dyson equation for the inverse propagator. If we return to eq. 1.18 and invert the propagator we discover the Schwinger–Dyson equation may also be written

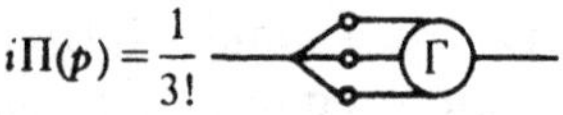

$$i\Pi(p) = \frac{1}{3!}$$

FIG. 31

The equation relates the exact propagator to the exact vertex. We can also obtain an equation for the vertex by differentiating 1.238 three times with respect to J and setting $J = 0$. Passing through the same steps which came from using the relations between Z and W, and W

and Γ, we can derive the Schwinger–Dyson equation for the vertex

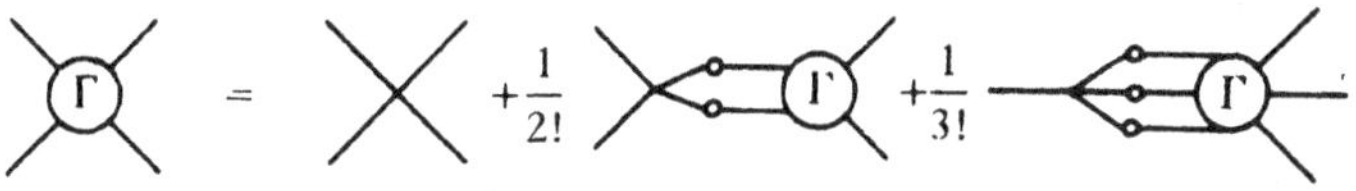

FIG. 32

It is an instructive exercise, which the reader should undertake, to go through the steps which lead to the equation for the vertex. Notice that the equation for the vertex as well as containing the vertex itself also contains the six-point proper vertex. This means that the two Schwinger–Dyson equations do not form a closed system of coupled equations. The six-point proper vertex, of course, always obeys its own equation, but this equation involves still more proper vertices not present in the previous equations. This is the difficulty of working with such a system of equations. A complete solution can only be accomplished by solving an infinite set of coupled equations. In perturbation theory the situation is less complex. This is because there are only two fundamental Green's functions: the bare propagator and the bare vertex. All the higher proper vertices may be defined in terms of these via Feynman diagrams.

There is no need to repeat the technique of functional Fourier transform for QED. After Fourier transforming with respect to ψ, $\bar\psi$ and A_μ and subsequent differentiation we obtain three equations. We show these in Fig. 33. The equation for the vertex in Fig. 34 was obtained by differentiating $Z[\zeta, \bar\zeta, J_\mu]$ first with respect to J_μ and then with respect to $\bar\zeta$ and ζ. We can also obtain an equation by

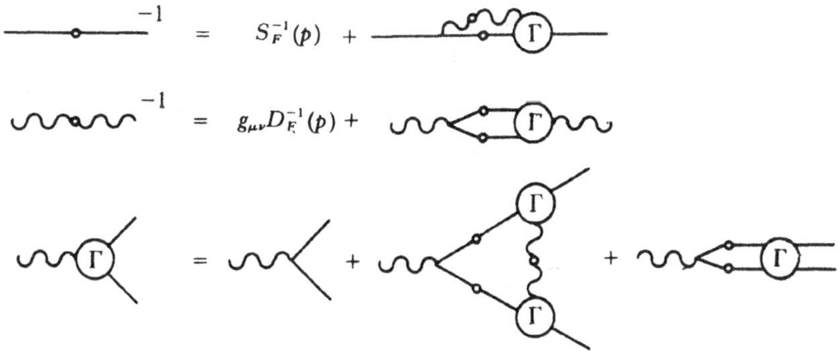

FIG. 33

differentiating with respect to J last rather than first. The equation is shown in Fig. 35.

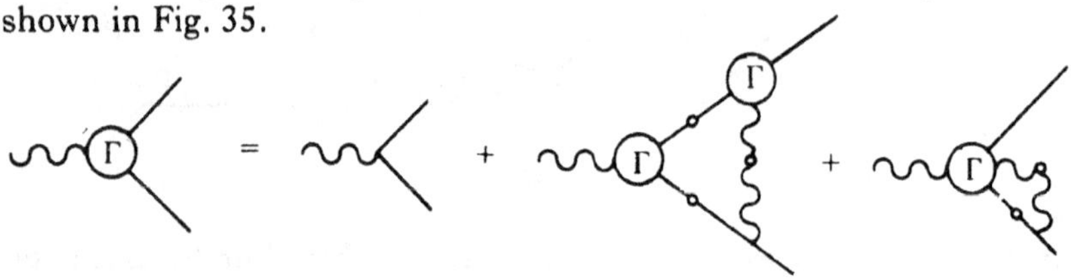

FIG. 34

These equations are basically the same although they may not look it. It is worth checking this by expanding the RHS of each equation into Feynman diagrams, and verifying that the same diagrams are generated by each equation. The Schwinger–Dyson equations for the electron and photon propagators may also be written in terms of the functions $\Sigma(p)$ and $D(p)$ introduced at the beginning of this chapter. In terms of these functions the equations become

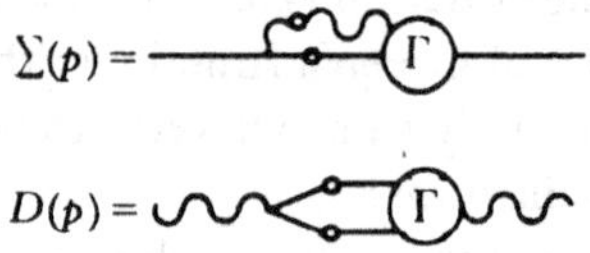

$$\Sigma(p) = \quad \text{———}$$

$$D(p) = \quad \text{———}$$

FIG. 35

This finishes our discussion of functional methods. These methods will be taken up again in Chapter 4.

References

1. J. Schwinger, *Phys. Rev.*, **91**, 713 (1953).
2. S. Weinberg, *Phys. Rev. Lett.*, **19**, 1264 (1967); A. Salam, *in* 'Proceedings of the Eighth Nobel Symposium' (N. Svartholm, ed.), Stockholm (1968).
3. J. Wess, B. Zumino, *Nucl. Phys.*, **B70**, 39 (1974), *Phys. Lett.*, **51B**, 239 (1974); A. Salam and J. Strathdee, *Nucl Phys.*, **B76**, 477 (1974).
4. R. P. Feynman and A. R. Hibbs, 'Quantum Mechanics and Path Integrals', McGraw-Hill.
5. N. Wiener, *J. Math. and Phys.*, **2**, 131 (1923).
6. I. M. Gel'fand and A. M. Yaglom, *J. Math. Phys.*, **1**, 48 (1960).
7. F. Riesz and B. Sz-Nagy, 'Functional Analysis', Frederick Ungar Publ. Co. (1955).
8. M. Kac, 'Probability and Related Topics in the Physical Sciences', Interscience (1959).

TWO

Dimensional Regularization and $\lambda\phi^4/4!$ Theory

Introduction

This chapter is devoted to the basic calculations of renormalization by the dimensional method. The reader is no doubt already aware of the enormous complexity of quantum field theory, and of the impracticality of presenting a complete argument conforming to the standards of rigour that many of us would like. The last chapter was concerned with formal considerations which pertain to all quantum field theories; this chapter will deal with two field theories—quantum electrodynamics and $\lambda\phi^4/4!$ theory—and present calculations and methods with the level of rigour relaxed so as to lessen the strain on the reader for whom the most relevant and important ideas are not obscured by the wealth of detail necessary in a completely rigorous discussion.

It is appropriate at the beginning of a discussion on renormalization to be aware that it is often loosely equated with the task of removing infinities from Feynman diagrams. This is in fact, not so and as we showed in the last chapter renormalization is necessary because of the presence of some interaction. However it should also be said that the bread and butter of calculations in renormalization theory do concern themselves with manipulating and ultimately removing infinite quantities. On the subject of infinities it is important to keep in mind the fact that although Feynman integrals are usually thought of as diverging at their unbounded upper limits of integration, i.e. ultraviolet divergence, in theories with zero mass particles, e.g. QED, there are also divergences coming from the origin in momentum space, infrared divergences. We shall have something to say about the infrared behaviour of QED in the next chapter. Another general point to realize is that although there exist general criteria such as Weinberg's theorem[1] for the finiteness of Feynman integrals, and prescriptions for extracting finite parts from those integrals which diverge, this is not the whole story. Suppose that once one has used Weinberg's theorem to discover which diagrams belonging to a given set are finite, and extracted the finite parts from the others according to ones renormalization prescription, one cannot yet conclude that the resulting theory, the so-called renormalized theory, is finite. A further theorem that requires proof before one can conclude that the renormalized theory is finite, is that which states that the renormalized perturbation expansion converges. In the absence of detailed information about the general properties of the nth term in the perturbation expansion it is rather difficult to decide on the validity of the above theorem. In general though it would be fair to say that it would take a courageous theorist to indicate himself on the side of convergence. Most theorists believe the expansion to be an asymptotic one at best.[2] In QED, for some value of the coupling constant, conditions have been described[2] under which the perturbation expansion will develop an essential singularity as a function of the coupling constant. Caution too is required when discussing the singularity structure in momentum space of scattering amplitudes in the context of perturbation theory. For the presence of a physical particle is usually signalled by the occurrence of a pole in the scattering amplitude and at this pole the scattering amplitude will not have a convergent perturbation expansion. Forearmed with this discussion of the various kinds of infinity that we can expect to encounter in perturbative quantum field theory, we hasten to add that we shall only be concerned in this book with those that arise in the individual Feynman diagrams themselves, and not with the deeper questions raised above. The organization of the rest of this chapter is as follows: we deal first with $\lambda\phi^4/4!$ theory and the structure of the infinities for some individual Feynman diagrams. We extract the finite parts by the conventional ultraviolet cut-off method, and then introduce the dimensional regularization method and compare the workings of the latter method with the former. We then go into some detail to explain how one constructs the analytic continuation in D, the dimension of space time, necessary for the dimensional method. The problem of the overlapping divergence is examined by this technique. We then show how to construct the inductive argument which leads to a proof of the finiteness, (subject to the qualifications made above), of the renormalized theory. We do not attempt the proof itself because of its length and complexity. However we do state the important theory of Weinberg and explain its significance.

Dimensional Regularization and $\lambda\phi^4/4!$ Theory

$\lambda\phi^4/4!$ theory is a theory of a single self-interacting scalar particle, and because there are no complications due to charge, internal symmetry, or spin, it is ideally suited to illustrate the properties of renormalization and the analytic structure of perturbative quantum field theory. Its use is limited to this pedagogical sphere since there appear, at the moment, to be no weakly or strongly coupled fundamental scalar particles.

The Lagrangian density is $L(x)$ where

$$L = \tfrac{1}{2}(\partial_\mu \phi)(\partial_\mu \phi) - \frac{m^2\phi^2}{2} - \frac{\lambda\phi^4}{4!}. \tag{2.1}$$

This gives rise to the Feynman rules already met in Chapter 1

internal line	——→—— p	$\dfrac{+i}{p^2 - m^2 - i\varepsilon}$
loop integration		$\displaystyle\int \frac{dk}{(2\pi)^4}$
vertex	$p+q+r+s=0$	$-i\lambda$
symmetry factors	S	

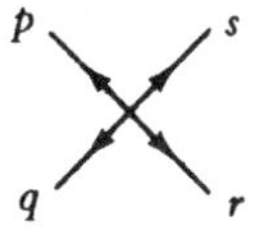

(2.2)

To illustrate the calculation of symmetry factors we give some examples

(a)   (b)

FIG. 1

In Fig. 1(a) $S = \tfrac{1}{6}$, Fig. 1(b) has $S = \tfrac{1}{2}$.

The reader who wishes to understand the necessity of the symmetry factors S in more detail should consider the diagram in configuration space rather than momentum space, and then compute S. We shall do this for one diagram, that of Fig. 1(a). We first label the vertices x and y as shown below, in Fig. 2

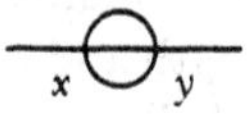

FIG. 2    FIG. 3

and then we disconnect the vertices of the diagram as shown in Fig. 3. The first task to be done is to count the number of ways that the two vertices of Fig. 3 can make the diagram of Fig. 2. The external line at x may be chosen in eight ways and then that at y in four ways. The remaining six lines may be joined in 3×2 ways to obtain the completed diagram. Thus the total number of ways is $8 \times 4 \times 6$. Finally we divide by 2! because the vertices x and y are identical and can be permuted, and the two permutational factors $(4!)^2$. The resultant symmetry factor is therefore

$$S = \frac{8 \times 4 \times 6}{(4!)^2 2!} = \frac{1}{6}. \tag{2.3}$$

Symmetry factors can easily be forgotten or incorrectly calculated. The best way to understand them is to work through several examples like the one above. There is the consolation that in more realistic theories with distinguishable particles, e.g. QED, these factors are not usually necessary. Another interesting property related to the combinatorics of the Lagrangian 2.1 is the selection rule that an odd number of ϕ's cannot scatter to give an even number of ϕ's or vice versa. One can easily prove this by an inductive argument where one tries to construct an arbitrary diagram with an odd number of external lines and proves that it is impossible. If one is more astute one can see that another proof can be constructed on noting that the result is due to the presence in the interaction term of the Lagrangian of an even rather than an odd power of the scalar field $\phi(x)$. This is in fact a consequence of an assumption often made that the Lagrangian should be symmetric under the interchange of ϕ and $-\phi$. The presence or absence of this symmetry has to do with spontaneous symmetry breaking, a very important topic, but one which we do not intend to go into here.

DIMENSIONAL REGULARIZATION AND $\lambda\phi^4/4!$ THEORY 65

We are now ready to calculate our first diagram which is shown with the momenta labelled in Fig. 4 below

$$p_1 + p_2 = p_3 + p_4$$

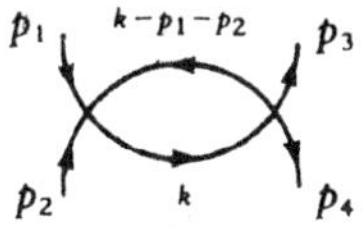

FIG. 4

The symmetry factor S is easily found to be $\frac{1}{2}$ and the Feynman integral is

$$\frac{1}{(2\pi)^4}\frac{\lambda^2}{2}\int dk \frac{1}{\{k^2 - m^2 + i\varepsilon\}\{(k - p_1 - p_2)^2 - m^2 + i\varepsilon\}}. \quad (2.4)$$

We introduce Feynman parameters α and β and use the identity,

$$\frac{1}{A_1 \ldots A_n} = \Gamma(n)\int_0^1 \frac{d\alpha_1 d\alpha_2 \ldots d\alpha_n \delta(\sum_{i=1}^n \alpha_i - 1)}{[\alpha_1 A_1 + \alpha_2 A_2 + \cdots \alpha_n A_n]^n} \quad (2.5)$$

where $\Gamma(n)$ is the usual gamma function which satisfies

$$\Gamma(n) = (n-1)! \quad (2.6)$$

for positive, integral n.

This yields the expression

$$\frac{\lambda^2 \Gamma(2)}{2(2\pi)^4}\int \frac{dk\, d\alpha\, d\beta\, \delta(\alpha + \beta - 1)}{[\alpha(k^2 - m^2 + i\varepsilon) + \beta\{(k - p_1 - p_2)^2 - m^2 + i\varepsilon\}]^2}$$

$$= \frac{\lambda^2}{2(2\pi)^4}\int \frac{dk\, d\alpha}{[\{k + (1-\alpha)(p_1 + p_2)\}^2 - m^2 + (1-\alpha)s - (1-\alpha)^2 s]^2} \quad (2.7)$$

where $s = (p_1 + p_2)^2$.

One is now tempted by the form of expression 2.7 to shift the origin of integration by changing variables from $k + (1-\alpha)(p_1 + p_2)$ to k. A note of caution should be sounded however, this shift of origin is only guaranteed to be valid if the integral is finite. This integral as we shall see later is in fact infinite; but nevertheless the origin shift is still allowed. An advantage of the dimensional regularization method which we shall introduce later is that it enables one to justify shifting the origin of integration because it is done at a stage when the integral is finite. With the origin shifted, 2.7 takes on the form

$$\frac{\lambda^2}{2(2\pi)^4} \int \frac{dk\, d\alpha}{[k^2+b^2]^2} \tag{2.8}$$

where

$$b^2 = -m^2 + \alpha(1-\alpha)s. \tag{2.9}$$

We may now do the loop momentum integration over k by using the formula

$$\int \frac{d^4k}{[k^2+b^2]^n} = i\pi^2 \frac{\Gamma(n-2)}{\Gamma(n)} \frac{1}{(b^2)^{n-2}}. \tag{2.10}$$

Unfortunately we notice that in 2.8 n has the value 2; and the gamma function in the numerator of 2.10 becomes infinite due to the presence of a pole, (the gamma function $\Gamma(n)$ has simple pole singularities when $n = 0, -1, -2, \ldots$). The source of this divergence can be traced to the unbounded upper limit of integration on the variable k. If we limit the magnitude of k in the integration to the value Λ, the so-called ultraviolet cut-off, it is then straightforward to estimate the growth of the integral as we remove this restriction and allow Λ to return to infinity. The result is that the integral has a dependence on Λ which makes it grow in proportion to $\ln(\Lambda^2/b^2)$; this is called in the literature a logarithmic divergence. In performing integrals, the energy axis of the 4-momentum k is rotated through $\pi/2$ in the complex energy plane. This has the effect of making the vector k Euclidean. This is done to eliminate the necessity of introducing discussions of conditional convergence in k^2, i.e. convergence which depends upon the sign of the variable k^2. For example the behaviours of the functions $\ln(-k^2/m^2)$ and $\exp(k^2/m^2)$ for large k^2 differ greatly if the sign of k^2 is changed. However, if we work with the rotation described above, k^2 is always negative. We regard this procedure as an analytic continuation to the region where k^2 is negative. We recover the values of expressions for positive k^2 by analytically continuing at the end of the calculation. This involves tacit assumptions about the analytic properties of perturbation theory which require proof. We cannot prove these assumptions here; we simply make them.

The task that we are now faced with is to extract a finite part from 2.8 and do away with the infinite part. We shall do this first by conventional means, using a cut-off Λ to make the integral finite, and then repeat the calculation using the dimensional regularization method.

Conventional Method

The knowledge that there is an infinity in the theory when the Lagrangian L is used leads us to the simple idea of modifying L so as to remove the infinity. In general the consequences of modifying L will be present in many more diagrams than Fig. 4, but for the moment we do not worry about this. We simply wish to see what we can learn from just working with Fig. 4. To this end we define the Lagrangian L' by

$$L' = \tfrac{1}{2}(\partial_\mu \phi)^2 - \frac{m^2\phi^2}{2} - \frac{\lambda\phi^4}{4!} + \frac{A\lambda\phi^4}{4!} = L + \frac{A\lambda\phi^4}{4!}. \qquad (2.11)$$

It is then trivial to see that the lowest order graph created by the new term $A(\lambda/4!)\phi^4$, (called a counter term) is simply that shown in Fig. 5

$$iA\lambda \equiv \quad \times$$

FIG. 5

where the bold dot indicates a vertex due to the counter term rather than the usual vertex. There are, of course, terms of higher order in A in the usual fashion but we shall not need these terms yet. To the order that we have worked out the complete four-point function is expressed by the diagrammatic equation

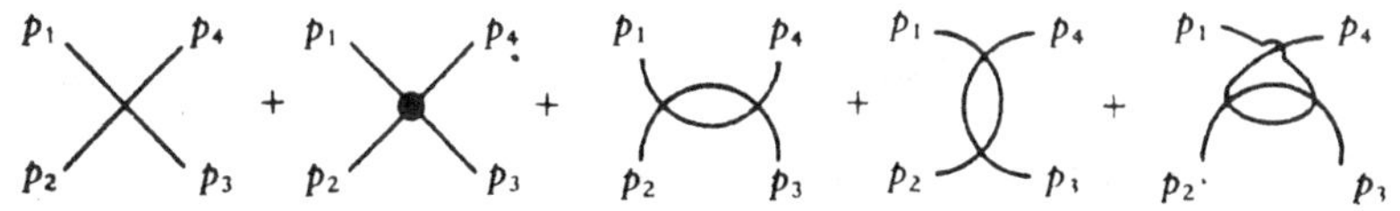

$$= -i\lambda + iA\lambda + \frac{\lambda^2}{2(2\pi)^4}\int \frac{dk\,d\alpha}{[k^2+b^2]^2} + \frac{\lambda^2}{2(2\pi)^4}\int \frac{dk\,d\alpha}{[k^2+c^2]^2} \times \frac{\lambda^2}{2(2\pi)^4}\int \frac{dk\,d\alpha}{[k^2+d^2]^2} \quad (2.12)$$

where c^2 and d^2 are functions constructed from b^2 by replacing s by t and u respectively, and t and u are the usual Mandelstam invariants so that

$$t = (p_1 - p_4)^2, \qquad u = (p_1 - p_3)^2. \qquad (2.13)$$

Of the three graphs in eq. 2.12 it is clearly only necessary to calculate one; we shall calculate graph No. 3. The procedure now begins by imposing a cut-off Λ^2 so that $k^2 \leq \Lambda^2$ in the integral over the variable k^2. This has the effect of making the integral in 2.8 finite but at the expense of introducing the parameter Λ^2; actually Λ^2 will not appear in the final version of the finite quantities, this to some extent compensates for having to introduce the parameter Λ^2.

Now we expand the integral in 2.8 about the point $s = 0$ and so obtain the equation

$$\frac{\lambda^2}{2(2\pi)^4} \int \frac{dk\, d\alpha}{[k^2+b^2]^2} = \frac{\lambda^2}{2(2\pi)^4} \int \frac{dk\, d\alpha}{[k^2-m^2+\alpha(1-\alpha)s]^2}$$

$$= \frac{\lambda^2}{2(2\pi)^4} \int \frac{dk\, d\alpha}{[k^2-m^2]^2}$$

$$+ \left\{ \frac{-\lambda^2}{(2\pi)^4} \int \frac{dk\, d\alpha\, \alpha(1-\alpha)}{[k^2-m^2+\alpha(1-\alpha)s]^3} \right\} s + \cdots = \Gamma(s). \quad (2.14)$$

The RHS of 2.14 can now be seen to have the property that only the first term is infinite in the limit when $\Lambda^2 \to \infty$. This is because the denominator in the integrand of each term after the first is of the form $[k^2-m^2+\alpha(1-\alpha)s]^n$; $n \geq 3$, if we use eq. 2.10 we see that for these values of n, the k space integral gives a finite answer. Confident that we have isolated the infinity we proceed to work on the first term on the RHS of eq. 2.14. To do this we first recall the expression for the element of volume in n-dimensional Euclidean space

$$d^n k = dk\, k^{n-1} \sin^{n-2}\theta_{n-1} \sin^{n-3}\theta_{n-2} \ldots d\theta_{n-1} \ldots d\theta_1 \quad (2.15)$$

where $0 \leq \theta_1 \leq 2\pi$ and all other angles range from 0 to π and $k = \sqrt{k^2}$. Using this formula when $n = 4$ we find that

$$\frac{\lambda^2}{2(2\pi)^4} \int^{\Lambda} \frac{dk\, d\alpha}{[k^2-m^2]^2} = \frac{i\lambda^2}{2(2\pi)^4} \frac{2\pi^2}{\Gamma(2)} \int_0^{\Lambda} \frac{dk\, k^3}{[k^2-m^2]^2}. \quad (2.16)$$

The α integration which was trivial has been done and the angular integration has been done by using the formula

$$\int d^n k\, f(k) = \frac{2\pi^{n/2}}{\Gamma(n/2)} \int dk\, k^{n-1} f(k)$$

valid in Euclidean space of n dimensions for a function which only depends on k. It is now a simple matter to evaluate the RHS of eq. 2.16 and obtain the result

$$\frac{\lambda^2}{2(2\pi)^4}\int^\Lambda \frac{dk\,d\alpha}{[k^2-m^2]^2} = \frac{i\lambda^2}{(2\pi)^4}\frac{\pi^2}{2}\left\{1+\frac{1}{(1-\Lambda^2/m^2)}+\ln\left(1-\frac{\Lambda^2}{m^2}\right)\right\}. \quad (2.17)$$

The logarithmic divergence is present as promised. The hitherto unspecified value of A is used to eliminate this Λ dependence by choosing

$$A = -\frac{\lambda\pi^2}{(2\pi)^4 2}\left\{\frac{1}{(1-\Lambda^2/m^2)}+\ln\left(1-\frac{\Lambda^2}{m^2}\right)\right\}. \quad (2.18)$$

We notice, of course, that when Λ is allowed to pass to infinity A becomes infinite so that the counter term is in fact infinite. However once the counter term has been used to render eq. 2.12 finite, the remaining finite part depends neither on Λ nor on the manner in which the limit $\Lambda \to \infty$ is taken. This is the only consolation that we have for having to introduce a formal counter term A which is infinite. Having made eq. 2.12 finite it remains to calculate this finite part; what we actually do is

 (i) calculate the expression $\partial\Gamma/\partial s$ from eq. 2.8
 (ii) integrate the result with respect to s to obtain $\Gamma(s)$.

There will be a constant of integration undetermined after performing this integration; this we may determine by noting that if we examine eq. 2.14 we see that our counter term A was chosen so that $\Gamma(s=0) = 0$. The careful reader will have noted that the choice of A is not unique, as the only requirement is that A must cancel the logarithmic divergence in Λ. This arbitrariness would allow $\Gamma(s=s_0)$ to be fixed to be some other value; this is perfectly legal and leads to a theory with the same physical content as the former. In fact the requirement that this be true is the starting point for the main topic of a later chapter; the renormalization group.

Differentiating Γ we find

$$\frac{\partial\Gamma}{\partial s} = \frac{-\lambda^2}{(2\pi)^4}\int \frac{dk\,d\alpha\,\alpha(1-\alpha)}{[k^2-m^2+\alpha(1-\alpha)s]^3}$$

$$= \frac{-i\pi^2\lambda^2}{(2\pi)^4\Gamma(3)}\int_0^1 \frac{d\alpha\,\alpha(1-\alpha)}{[\alpha(1-\alpha)s-m^2]} \quad (2.19)$$

using the formula of eq. 2.10. The integration over α is best done after first integrating with respect to s. (We may freely change the order of these analytic operations since the resulting expression is finite). The expression to evaluate is then

$$\Gamma(s) = \frac{-i\lambda^2 \pi^2}{2(2\pi)^4} \int_0^1 d\alpha \ln[\alpha(1-\alpha)s - m^2] + C \qquad (2.20)$$

with C the constant of integration. We then find after performing the final integration that

$$\Gamma(s) = \frac{-i\lambda^2 \pi^2}{2(2\pi)^4} \left[-2 + 2(\tfrac{1}{4} - m^2/s)^{1/2} \ln\left\{ \frac{[\tfrac{1}{2} + (\tfrac{1}{4} - m^2/s)^{1/2}]}{[\tfrac{1}{2} - (\tfrac{1}{4} - m^2/s)^{1/2}]} \right\} + C \right]. \qquad (2.21)$$

We then fix C by the particular normalization that is desired.

The inclusion of the remaining diagrams gives the final expression

$$= -i\lambda + \Gamma(s) + \Gamma(t) + \Gamma(u). \qquad (2.22)$$

We point out in passing that the threshold square root branch cut at $s = 4m^2$ which is required by unitarity[3] is present. So much for the cut-off method of extracting finite parts. Before going on to the details of the definitions of renormalization constants in terms of counter terms, we go straight on to repeat the above calculation by the dimensional method.

The Dimensional Method[4]

We begin with the basic integral of 2.8 and consider it to be an integral over a D-dimensional loop momentum k rather than a four-dimensional momentum. D is a variable which is allowed to vary continuously and need not be an integer, a rational or a real etc. It is important to understand that this need not involve us in some bizarre mathematics where one has to define non-integral numbers of variables or the like. The sense in which the integral is to be regarded as a D-dimensional one is the sense in which the integral may be defined as a function of D by the process of analytic continuation in D. To make this clear consider the generalization of the formula of eq. 2.10 which is valid in n-dimensional space

$$\int \frac{d^n k}{[k^2+b^2]^p} = i\pi^{n/2} \frac{\Gamma(p-n/2)}{\Gamma(p)} \frac{1}{(b^2)^{p-n/2}}. \quad (2.23)$$

This formula is defined, by analytic continuation of the RHS, to have a value when $n = D$ where D is not restricted to integral values. It transpires that this is all one needs to develop expressions for 'D-dimensional' Feynman integrals. We shall discuss the construction of this analytic continuation in a bit more detail after this example. Returning to the basic integral 2.8 but now defining it in D dimensions and using the formula of eq. 2.23 we find that

$$\frac{\lambda^2}{2(2\pi)^4} \int \frac{d^D k \, d\alpha}{[k^2+b^2]^2} = \frac{i\lambda^2 \pi^{D/2}}{2(2\pi)} \frac{\Gamma(2-D/2)}{\Gamma(2)} \int_0^1 \frac{d\alpha}{(b^2)^{2-D/2}}. \quad (2.24)$$

The striking property of eq. 2.24 is that it is perfectly finite if D is chosen to be slightly less than 4, say $D = 4-\varepsilon$, $\varepsilon > 0$. Indeed infinities are seen only to arise when the gamma function $\Gamma(2-D/2)$ is infinite. This, according to the well-known properties of Γ, is when $2-D/2$ is zero or a negative integer, when Γ has simple poles. This remarkable formula allows us to separate 2.24 at once into those parts which will be infinite and those which will be finite at the value $D = 4$. To do this we require the expressions[†]

$$\Gamma(x) \underset{x\to 0}{\to} \frac{1}{x} - \gamma$$

$$\frac{1}{(b^2)^{2-D/2}} = 1 - (2-D/2) \ln b^2 + \cdots. \quad (2.25)$$

Thus when $D/2 - 2 \approx 0$ we may replace the RHS of 2.24 by

$$\frac{i\pi^2 \lambda^2}{2(2\pi)^4} \left\{ \frac{1}{2-D/2} - \gamma \right\} \int_0^1 d\alpha \, \{1 - (2-D/2) \ln b^2 + \cdots\}. \quad (2.26)$$

The infinite part is now a simple pole in D as $D \to 4$ with residue $-i\pi^2 \lambda^2/(2\pi)^4$, while the finite part is

$$-\frac{i\pi^2 \lambda^2}{2(2\pi)^4} \left\{ \int_0^1 d\alpha \, \ln b^2 + \gamma \right\}. \quad (2.27)$$

[†] γ is Euler's constant and is given by $\gamma = \lim_{n\to\infty} (1 + \frac{1}{2} + \cdots + \frac{1}{n} - \ln n) = 0.577\ldots$.

To eliminate the infinite part we return to the Lagrangian $L' = L + (A\lambda/4!)\phi^4$. All that is necessary to render everything finite is to choose† $A = -3\pi^2\lambda/(2\pi)^4(D-4)$. A is an infinite quantity when $D = 4$ as it was in the cut-off method; however its evaluation is a much swifter and simpler process than in the previous cut-off method. Also there has been no loss of Lorentz invariance at any stage during the calculation as there can be in some cut-off methods. As for the finite part we see by inspection that apart from a constant term it is precisely the same as the expression obtained above. To see where the arbitrariness of the constant term enters in the dimensional method, we make the observation that the LHS of 2.24 may always be multiplied by $(C)^{D-4}$ where C is a constant, since this does not alter the value of the integral in the physical number of dimensions. However reference to eq. 2.26 shows that this multiplicative factor when expanded as a series in $(D-4)\ln C$ will contribute an amount $-i\pi^2\lambda^2 \ln C/(2\pi)^4$ to the finite part. This is the arbitrariness of the constant term which we discovered during the first calculation. Of course, if it were not for the presence of the infinity, multiplication by $(C)^{D-4}$ would not alter the value of the Feynman integral. It is precisely when an infinity is present that care must be taken about a lack of uniqueness that may occur and go unnoticed. The dimensional method may be a little perplexing at first if the reader is used to a more conventional method. It is nevertheless a superior method since it is quicker and more elegant; also it may be applied to show that the non-Abelian gauge theories are renormalizable. Indeed it was not until the invention of this method that the renormalizability of non-Abelian gauge theories was proved.[5] To overcome this perplexing feeling the recommendation is clear: the student should work through several examples using both methods.

To make the discussion of the dimensional method complete one has to show how one can construct the desired analytic continuation. We divide this discussion into two parts, single loop and multiloop diagrams.

Single Loop Diagrams

There are two cases to consider, continuation to $D < 4$, and continuation to $D > 4$; the continuation will, of course exist for D complex but

---

† The factor 3 comes from the fact that there are three diagrams each with the same infinite part.

we shall not exploit this. The first thing to make clear is what we mean by a D-dimensional vector in Minkowski space in the cases where D is an integer. If k is a vector in four-dimensional Minkowski space and has the representation

$$k = (k_0, k_1, k_2, k_3), \qquad k^2 = k_0^2 - k_1^2 - k_2^2 - k_3^2. \tag{2.28}$$

Then a D-dimensional k is obtained by leaving the energy component as it stands and extending only the spatial part so that in D (integral) dimensions k has the representation

$$k = (k_0, k_1, \ldots k_{D-1}), \qquad k^2 = k_0^2 - k_1^2 \ldots k_{D-1}^2$$
$$= k_0^2 - r^2. \tag{2.29}$$

It should be pointed out that we do not employ this D-dimensional representation of momenta for the momenta on the external lines; its use is restricted to the internal loop momenta. Taking the integral of 2.24 and using 2.29 we have the integral

$$\frac{\lambda^2}{2(2\pi)^4} \int_{-\infty}^{\infty} dk_0 \int d^{D-1}\Omega \int_0^{\infty} dr\, r^{D-2} \int_0^1 \frac{d\alpha}{[k_0^2 - r^2 + b^2]^2} \tag{2.30}$$

$d^{D-1}\Omega$ is the element of solid angle in $D-1$ dimensions and is easily seen to provide a factor $2\pi^{(D-1)/2}/\Gamma[(D-1)/2]$ and the integral becomes

$$\frac{\lambda^2}{(2\pi)^4} \frac{\pi^{(D-1)/2}}{\Gamma[(D-1)/2]} \int_{-\infty}^{\infty} dk_0 \int_0^{\infty} dr\, r^{D-2} \int_0^1 \frac{d\alpha}{[k_0^2 - r^2 + b^2]^2}. \tag{2.31}$$

This integral exists for $1 < D < 4$, at $D = 4$ it has a divergence in the form of a simple pole as we saw in the last section. Moreover at $D = 1$ both the gamma function and the integral become infinite, while for $D < 1$ the integral diverges due to its behaviour at the lower limit of integration. To analytically continue this integral we use the device of performing integration by parts, but it is convenient to do this with respect to the variable r^2 rather than r. If we integrate once we derive the formula

$$\frac{\lambda^2}{2(2\pi)^4} \frac{\pi^{(D-1)/2}}{\Gamma(D-1/2)} \int_{-\infty}^{\infty} dk_0 \int_0^{\infty} dr^2\, (r^2)^{(D-3)/2} \int_0^1 \frac{d\alpha}{[k_0^2 - r^2 + b^2]^2}$$
$$= \frac{\lambda^2}{2(2\pi)^4} \frac{\pi^{(D-1)/2}}{\Gamma[(D+1)/2]} \int_{-\infty}^{\infty} dk_0 \int_0^{\infty} dr^2\, (r^2)^{(D-1)/2} \int_0^1 \frac{d\alpha\,(-4r)}{[k_0^2 - r^2 + b^2]^3}. \tag{2.32}$$

This formula requires some explanation. The boundary term is zero for $D<5$; the remaining term comes from using the identity

$$dr\, r^{D-2} = \frac{1}{2} dr^2 \frac{2}{D-1} \frac{d}{dr^2}(r^2)^{(D-1)/2} \quad (2.33)$$

to perform the integration by parts. If we inspect 2.32 we see that the RHS exists not for $1<D<4$ but for $-1<D\leq 4$. This expression must be the analytic continuation we require because it is equal to the original integral in the region $1<D<4$ and is analytic in the larger region $-1<D<4$. To enlarge the domain of the analytic continuation it is only necessary to repeat the process. The general formula which results from integration by parts n times is given below

$$\frac{\lambda^2}{2(2\pi)^4} \frac{2\pi^{(D-1)/2}}{\Gamma[(D+2n-1)/2]} \int_{-\infty}^{\infty} dk_0 \int_0^{\infty} dr^2\, (r^2)^{D+2n-1}\left(-\frac{d}{dr^2}\right)^n$$

$$\cdot \int_0^1 \frac{d\alpha}{[k_0^2 - r^2 + b^2]^2}. \quad (2.34)$$

The domain of analyticity of 2.34 is, $1-2n<D<4$, so that by taking n large enough D may be made as small as we like. This disposes of the problem of continuing D to smaller values. We now wish to be able to continue to values of D larger than four. The same general method is used. We start with the integral of 2.31 and begin integration by parts, there is a difference here though in that we wish to integrate both with respect to k_0 and r. To make it easier to see how the process works we draw attention to the identity

$$1 = \frac{1}{2}\left(\frac{dk_0}{dk_0} + \frac{dr}{dr}\right) \quad (2.35)$$

which we have used in the integrand, a trick familiar in integration by parts. Performing the process once yields the equation

$$I = \frac{\lambda^2}{(2\pi)^4} \frac{\pi^{(D-1)/2}}{\Gamma[(D-1)/2]}\left[\int_{-\infty}^{\infty} dk_0 \int_0^{\infty} dr\, r^{D-2} \frac{(-2k_0^2 + 2r^2)}{[k_0^2 - r^2 + b^2]^3}\right.$$

$$\left.+\frac{1}{2}\int_{-\infty}^{\infty} dk_0 \int_0^{\infty} dr\, r^{D-2} \frac{(D-2)}{[k_0^2 - r^2 + b^2]^2}\right] = I_C + \frac{D-2}{2} I \quad (2.36)$$

where I is the integral of 2.31 and I_c has the obvious definition from the above equation. We find then that

$$I = -\frac{1}{D-4} I_c. \tag{2.37}$$

We have found the pole at $D=4$ and constructed the analytic continuation I_c for $D>4$. I_c is finite for $1<D<5$. It should now be clear that as in the previous case we may integrate by parts as many times as we wish to construct the analytic continuation I_c of I for arbitrarily large values of D. To complete the discussion we must deal with the multi-loop diagrams.

Multi-loop Diagrams

The construction of the analytic continuation in this case is quite complicated algebraically speaking but contains no new idea or concept, simply more work. For this reason, rather than construct the analytic continuation for a general number p of loops, we shall show how to construct it for two loops. It is a straightforward, if tedious, matter to derive the form of the continuation for p loops, and we shall give the formulae for p loops as a goal towards which the industrious reader may aim. As we did for one loop diagrams we begin with the continuation to smaller values of D. Let the two internal loop momenta be k_1 and k_2. The integration measure is divided as shown below

$$d^D k_1 \, d^D k_2 \mapsto d^4 k_1' \, d^4 k_2' \, d^{D-4} k_1 \, d^{D-4} k_2 \tag{2.38}$$

where k_1 and k_2 have been split into four vectors, two of dimension four, k_1', k_2' and two of dimension $D-4$. The angular integration yields the equation

$$\int d^D k_1 \, d^D k_2 \mapsto \frac{2^2 \pi^{(2D-9)/2}}{\Gamma\left(\frac{D-4}{2}\right)\Gamma\left(\frac{D-5}{2}\right)} \int d^4 k_1' \int d^4 k_2' \int_0^\infty dr_1 \, r_1^{D-5}$$

$$\cdot \int_{-\infty}^\infty dr_5^2 \int_0^\infty dr_2 \, r_2^{D-6}. \tag{2.39}$$

We have chosen the fifth axis in the k_2 space in the direction of the $(D-4)$-dimensional vector k_1. It is possible to write 2.39 in a more compact form by defining two new two-dimensional vectors p_1 and p_2 by

$$p_1 = \begin{pmatrix} r_1 \\ 0 \end{pmatrix}, \qquad p_2 = \begin{pmatrix} r_5^2 \\ r_2 \end{pmatrix}. \tag{2.40}$$

We may now write

$$\int d^D k_1 \, d^D k_2 \mapsto \frac{2\pi^{(2D-11)/2}}{\Gamma\left(\frac{D-4}{2}\right)\Gamma\left(\frac{D-5}{2}\right)}$$
$$\cdot \int d^4 p_1 \int d^4 p_2 \int d^2 p_1 \int d^2 p_2 \, (\varepsilon^{ab} p_1^a p_2^b)^{D-6} \theta(\varepsilon^{ab} p_1^a p_2^b) \quad (2.41)$$

ε_{ab} is the permutation symbol in two dimensions and θ is the step function which satisfies

$$\theta(x) = \begin{cases} 1 & x > 0 \\ 0 & x < 0. \end{cases} \quad (2.42)$$

The θ function is needed to impose the lower limit of zero in the r integration. The passage from 2.40 to 2.41 follows immediately from the properties of θ and ε_{ab} and the definitions of the two-dimensional integration measures for p_1 and p_2. The analytic continuation is constructed, as usual, by repeated integration by parts with respect to the variables p_1 and p_2. If we define I by

$$I = \int d^D k_1 \, d^D k_2 \, F(k_1^2, k_2^2, (k_1+k_2)^2) \quad (2.43)$$

then the continuation I_c to smaller values of D is given by

$$I_c = \frac{2\pi^{(D-11)/2}}{\Gamma\left(\frac{D-4+2n}{2}\right)\Gamma\left(\frac{D-5+2n}{2}\right)}$$
$$\cdot \int d^4 k_1' \, d^4 k_2' \int d^2 p_1 \, d^2 p_2 \, (\varepsilon^{ab} p_1^a p_2^b)^{D-6+2n}$$
$$\cdot \theta(\varepsilon^{ab} p_1^a p_2^b) \left\{ \frac{\partial^2}{\partial p_1^2 \partial p_2^2} + \frac{\partial^2}{\partial p_1^2 \partial (p_1+p_2)^2} + \frac{\partial^2}{\partial p_2^2 \partial (p_1+p_2)^2} \right\}^n F. \quad (2.44)$$

In the above the integration by parts has been performed n times with the aid of the formula

$$(\varepsilon^{ab} p_1^a p_2^b)^p = \frac{1}{(p+1)(p+2)} \varepsilon_{ab} \frac{\partial}{\partial p_1^a} \frac{\partial}{\partial p_2^b} (\varepsilon_{cd} p_1^c p_2^d)^{p+1}. \quad (2.45)$$

We now give the formula that is the analogue of 2.44 for a multi-loop diagram with l loop momenta $k_1, \ldots k_l$. It is

$$I_c = 2 \prod_{j=1}^{p} \pi^{(D-j-4)/2} \frac{\Gamma\left(\frac{p-j+1}{2}\right)}{\Gamma\left(\frac{D-3-j+2n}{2}\right)} \int d^4k_1' \ldots d^4k_l' \int d^k p_1 \ldots d^k p_l$$

$$\cdot \theta(\det p)(\det p)^{D-4+l+2n} \left[\det\left(\frac{-1}{2}\frac{\partial}{\partial a_{ij}} \frac{-1}{2}\frac{\partial}{\partial a_{ji}}\right)\right]^{2n} F(k). \quad (2.46)$$

In the above we have introduced the vectors $p_1, \ldots p_p$ where

$$p_1 = \begin{pmatrix} r_1 \\ 0 \\ \vdots \\ 0 \end{pmatrix}, \quad p_2 = \begin{pmatrix} (k_1)_1 \\ r_2 \\ 0 \\ \vdots \\ 0 \end{pmatrix} \cdots \quad p_p = \begin{pmatrix} (k_1)_1 \\ \vdots \\ (k_1)_{l-1} \\ r_l \end{pmatrix} \quad (2.47)$$

and made use of the formulas

$$\varepsilon_{a_1 \ldots a_l} p_1^{a_1} \ldots p_l^{a_l} = \det p$$

$$(\det p)^n = \frac{1}{(n+1) \ldots (n+l)} \left(\det \frac{\partial}{\partial p}\right) (\det p)^{n+1} \quad (2.48)$$

and the definition

$$a_{ij} = p_i \cdot p_j. \quad (2.49)$$

To construct the continuation for larger values of D than four we extend the identity 2.35 and use the identity

$$1 = \sum_{i=1}^{4} \sum_{j=1}^{l} \frac{dp_{ji}'}{dp_{ji}'} + \frac{1}{l^2} \sum_{i=1}^{l} \sum_{j=1}^{l} \frac{dp_{ji}}{dp_{ji}}. \quad (2.50)$$

The integration by parts is now performed and the formula obtained has the form

$$I = \frac{1}{Dl - n} I_c. \quad (2.51)$$

Equation 2.51 has the property that n, which is determined by counting powers of momentum in the integrand, tells us the nature of the divergence of I when $D = 4$. If $4l - n = 2, 1, 0, \ldots$ the divergences will be quadratic, linear and logarithmic respectively.

The Overlapping Divergence

This completes our discussion of the analytic continuation. We now wish to use the technique to examine another basic diagram in $\lambda\phi^4/4!$ theory. The diagram in Fig. 6 is the lowest order, non-trivial diagram

FIG. 6

contributing to the propagator. The symmetry factor for this diagram is $\frac{1}{6}$ and the integral to be examined is

$$I = \frac{i}{6}\frac{\lambda^2}{(2\pi)^8} \int \frac{dk\, dl}{[k^2-m^2][l^2-m^2][(p+l-k)^2-m^2]} \quad (2.52)$$

Inspection of I shows that it is divergent. There are various regions of l and k in which the integral diverges, e.g. l small, k large, k small and l large, these give rise to logarithmic divergences. But if we take both k and l large the integral diverges quadratically. Furthermore this is what is called an overlapping divergence because one cannot ascribe the divergence to any subdiagram of Fig. 6. Also there does not exist a change of variables from l, k to l', k' for which the divergence only appears when one does only one of the loop integrations l' and k'. Overlapping divergences are much more difficult to deal with than the simpler kind (i.e. just associated with one loop of a diagram). They have the property that if one uses the Feynman parameters of eq. 2.5 and the formula of eq. 2.10 to evaluate the integral the divergence moves in part from the loop-momenta integrations to the integrations over the Feynman parameters. This usually makes the dimensional method very difficult to use and caution must be exercised in the face of multiloop calculations. Now as implied above not only does the overall diagram of Fig. 6 diverge but so do the various subdiagrams of which it is made up. We shall need therefore counter terms to eliminate the divergences in the subdiagrams as well as a counter term to eliminate the divergence due to the whole diagram. The subdiagrams are shown in Fig. 7(a)–(c). The subdiagrams of Fig. 7 can easily be checked to have logarithmic divergences. This reasoning leads one to believe that four counter terms are needed to eliminate these four

DIMENSIONAL REGULARIZATION AND $\lambda\phi^4/4!$ THEORY

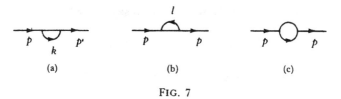

FIG. 7

sources of divergence. This is indeed the case, and is the proper way to tackle the overlapping divergence. The origins of the four subtractions that must be made from I can be isolated very simply in the dimensional method. They are as follows

(a) a subtraction due to a divergence exposed by integrating by parts with respect to k;
(b) a subtraction due to a divergence exposed by integrating by parts with respect to l;
(c) a subtraction due to a divergence exposed by changing variables from l to $k-l$ and integrating by parts with respect to k;
(d) a subtraction due to a divergence exposed by using the identity

$$1 = \frac{1}{2D} \sum_{i=1}^{D} \left(\frac{\partial k_i}{\partial k_i} + \frac{\partial l_i}{\partial l_i}\right) \quad (2.53)$$

in the integrand.

Taking the operation described in (d) first and using the method that we have just developed we find

$$I = \frac{1}{2D-6} I_c,$$

$$I_c = +\frac{i\lambda^2}{6(2\pi)^8} \int dk \, dl \left[\frac{2m^2}{[k^2-m^2]^2[l^2-m^2][(p+l-k)^2-m^2]} \right.$$

$$+ \frac{2m^2}{[k^2-m^2][l^2-m^2]^2[(p+l-k)^2-m^2]}$$

$$\left. + \frac{-2p^2+2m^2+2p\cdot(k-l)}{[k^2-m^2][l^2-m^2][(p+l-k)^2-m^2]^2} \right]. \quad (2.54)$$

When $D=4$, $2D-6=2$ verifying the presence of the quadratic divergence already mentioned. For the other three diagrams we introduce the notation of putting a cross in the subdiagram to denote the counter term. The counter terms of (a), (b) and (c) can then be given as shown in Fig. 8. Now we need to say a word or two about the structure of the divergences in these subdiagrams. Consider (a) without the cross; integration over k will yield a pole at $D=4$ with a constant residue (the residue will be a dimensionless constant since the divergence is logarithmic); however we have still to integrate over l.

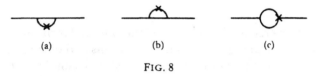

FIG. 8

Integration over the parts which contribute a finite quantity will contribute to the residue so that terms like

$$\frac{\ln m^2}{D-4} \qquad (2.55)$$

will appear. Such terms cannot be allowed in the residues of poles because they are not polynomials in the external momenta and when counter terms are introduced in the Lagrangian to eliminate them they in general violate locality. Hence in the sum of all counter terms such terms must vanish. This is so, terms of the form 2.55 from (a)...(c) cancel against similar terms coming from the diagram as a whole in 2.54. This is the making precise of the notion that the various divergences overlap. The subtraction denoted by the subdiagram of Fig. 8(a) is

$$I_{(a)} = \frac{i\lambda^2}{6(2\pi)^8} \int d^D l\, P \int \frac{d^D k}{[k^2-m^2][l^2-m^2][(p+l-k)^2-m^2]}. \qquad (2.56)$$

The symbol P denotes that a projection has been made on to the part of what follows which contains a pole in D in the physical number of dimensions, i.e. four. This pole part we can calculate by the method used for the one loop diagram and obtain

$$I_{(a)} = \frac{2\lambda^2 \pi^2}{6(2\pi)^8(D-4)} \int \frac{d^D l}{[l^2-m^2]}. \qquad (2.57)$$

This expression we integrate by parts with respect to l and find

$$I_{(a)} = \frac{2\lambda^2 \pi^2}{6(2\pi)^8(D-4)(D-2)} \int \frac{d^D l\, 2m^2}{[l^2-m^2]^2}. \qquad (2.58)$$

DIMENSIONAL REGULARIZATION AND $\lambda\phi^4/4!$ THEORY

The symmetry of the diagram means that $I_{(a)} = I_{(b)}$ which we write for convenience as

$$I_{(b)} = \frac{2\lambda^2\pi^2}{6(2\pi)^8(D-4)(D-2)} \int \frac{d^D k\, 2m^2}{[k^2-m^2]^2}. \qquad (2.59)$$

The third subtraction term $I_{(c)}$ is also equal to $I_{(a)}$ and we write it for convenience as

$$I_{(c)} = \frac{2\lambda^2\pi^2}{6(2\pi)^8(D-4)(D-2)} \int \frac{d^D l\, \{2m^2 - 2p^2 + 2(k-l)\cdot p\}}{[(p+l-k)^2 - m^2]^2}. \qquad (2.60)$$

The next thing to do is to add the subtraction terms $I_{(a)}$, $I_{(b)}$ and $I_{(c)}$ to the integral I given in eq. 2.54. We obtain

$$I + I_{(a)} + I_{(b)} + I_{(c)} = \frac{+i\lambda^2}{(2D-6)6(2\pi)^8}$$

$$\cdot \Bigg[\Bigg\{ \int \frac{dk\, dl\, 2m^2}{[k^2-m^2]^2[l^2-m^2][(p+l-k)^2-m^2]} + \frac{2i\pi^2(2D-6)}{(D-4)(D-2)}$$

$$\cdot \int \frac{dk\, 2m^2}{[k^2-m^2]^2} \Bigg\} + \Bigg\{ \int \frac{dk\, dl\, 2m^2}{[k^2-m^2][l^2-m^2]^2[(p+l-k)^2-m^2]}$$

$$+ \frac{2i\pi^2(2D-6)}{(D-4)(D-2)} \int \frac{dl\, 2m^2}{[l^2-m^2]^2} \Bigg\}$$

$$+ \Bigg\{ \int \frac{dk\, dl\, 2m^2}{[k^2-m^2][l^2-m^2][(p+l-k)^2-m^2]^2}$$

$$+ \frac{2i\pi^2(2D-6)}{(D-4)(D-2)} \int \frac{dl\, (2m^2 - 2p^2 + 2p\cdot(k-l))}{[(p+l-k)^2-m^2]^2} \Bigg\} \Bigg]. \qquad (2.61)$$

We have grouped the terms together to show how the subtractions operate. It is now possible to see how the dimensional method and its analytic continuation separate the sources of overlapping divergence in the whole diagram and show how they cancel against the divergences of the subdiagrams. The expression in 2.61 has not yet received a subtraction for the quadratic divergence of the whole diagram, but it is guaranteed not to have any poles with $\ln m^2$ residues which we wished to avoid. The way to verify this is by explicit calculation which we now explain. The three groupings of terms in 2.61 are all the same so it suffices to consider one of them. We shall consider the first one.

What we shall do is to transform the term so that its infinities in the form of poles at $D=4$ are explicit; and thus, as it happens, also find the finite part. The expression that we are working with is

$$A = \int \frac{dk}{[k^2 - m^2]^2} \left\{ \frac{\int dl}{[l^2 - m^2][(p+l-k)^2 - m^2]} - \frac{2i\pi^2(2D-6)}{(D-4)(D-2)} \right\}. \quad (2.62)$$

If we combine the two single propagators by the Feynman identity and perform the l integration in the now familiar fashion; then

$$A = i\pi^2 \int d^D k \left[\left[\int_0^1 \frac{d\alpha\, \Gamma(2-D/2)}{\{\alpha(1-\alpha)(k-p)^2 - m^2\}^{2-D/2}} - \frac{2(2D-6)}{(D-4)(D-2)} \right] \right.$$
$$\left. \cdot \frac{1}{[k^2 - m^2]^2} \right.. \quad (2.63)$$

To tackle the first term in 2.63 we use a generalization of the Feynman identity 2.5

$$\frac{1}{A_1^n A_2^m} = \frac{\Gamma(n+m)}{\Gamma(n)\Gamma(m)} \int_0^1 \frac{d\alpha\, d\beta\, \delta(\alpha+\beta-1)\alpha^{n-1}\beta^{m-1}}{[\alpha A_1 + \beta A_2]^{n+m}}. \quad (2.64)$$

The result of this is to cast 2.63 into the form

$$-\pi^4 \left[\frac{\Gamma\left(\frac{4-D}{2}\right)}{\Gamma\left(\frac{6-D}{2}\right)} \int_0^1 d\beta (1-\beta)^{(4-D)/2} \int_0^1 \frac{d\alpha}{\{\alpha(1-\alpha)\}^{(4-D)/2}} \right.$$

$$\cdot \left\{ \frac{\Gamma(5-D)\left\{ -\frac{m^2}{\alpha(1-\alpha)} + p^2 \beta^2 \right\}}{\left\{ (p^2 - m^2)\beta - \frac{m^2(1-\beta)}{\alpha(1-\alpha)} - p^2\beta^2 \right\}^{5-D}} \right.$$

$$\left. + \frac{\Gamma(4-D)D/2}{\left\{ (p^2 - m^2)\beta - \frac{m^2(1-\beta)}{\alpha(1-\alpha)} - p^2\beta^2 \right\}^{4-D}} \right\}$$

$$\left. - \frac{2(2D-6)}{(D-4)(D-2)} \cdot \frac{\Gamma\left(\frac{4-D}{2}\right)}{(-m^2)^{2-D/2}} \right]. \quad (2.65)$$

It is now possible to read off the sources of poles; there are four such sources. These are, the function $\Gamma(4-D)/2$ which occurs twice in different places, the function $\Gamma(4-D)$, and the explicit pole $1/(D-4)$. It is now a straightforward matter to expand 2.65 about $D=4$ and verify the absence of poles with residues proportional to $\ln m^2$. Inspection will also show the presence of double poles for which counter terms are needed. Thus the final counter term which we need will be a linear combination of a double and a single pole at $D=4$. The infinite parts add up finally in the total expression which makes up I to an expression which we shall deal with in a moment. The finite part of I is

$$\frac{+i\lambda^2}{2(16\pi^2)^2}\left[-\gamma\int_0^1 d\beta\,\beta\int_0^1 \frac{d\alpha\left\{-\frac{m^2}{\alpha(1-\alpha)}+p^2\beta\right\}m^2}{\left\{(p^2-m^2)\beta-\frac{m^2(1-\beta)}{\alpha(1-\alpha)}-p^2\beta^2\right\}}\right.$$

$$-2\int_0^1 d\beta\,\beta\ln(1-\beta)\int_0^1 \frac{d\alpha\left\{-\frac{m^2}{\alpha(1-\alpha)}+p^2\beta\right\}m^2}{\left\{(p^2-m^2)\beta-\frac{m^2(1-\beta)}{\alpha(1-\alpha)}-p^2\beta^2\right\}}$$

$$+\int_0^1 d\beta\,\beta\int_0^1 d\alpha\,\ln\{\alpha(1-\alpha)\}\frac{\left\{-\frac{m^2}{\alpha(1-\alpha)}+p^2\beta\right\}m^2}{\left\{(p^2-m^2)\beta-\frac{m^2(1-\beta)}{\alpha(1-\alpha)}-p^2\beta^2\right\}}$$

$$-m^2\int_0^1 d\beta\,\beta\int_0^1 d\alpha\,\ln^2\left\{(p^2-m^2)\beta-\frac{m^2(1-\beta)}{\alpha(1-\alpha)}-p^2\beta^2\right\}$$

$$+3\gamma m^2\int_0^1 d\beta\,\beta\int_0^1 d\alpha\,\ln\left\{(p^2-m^2)\beta-\frac{m^2(1-\beta)}{\alpha(1-\alpha)}-p^2\beta^2\right\}$$

$$+2m^2\int_0^1 d\beta\,\beta\int_0^1 d\alpha\left\{-\frac{m^2}{\alpha(1-\alpha)}+p^2\beta^2\right\}$$

$$\left.\cdot\ln\frac{\left\{(p^2-m^2)\beta-\frac{m^2(1-\beta)}{\alpha(1-\alpha)}-p^2\beta^2\right\}}{\left\{(p^2-m^2)\beta-\frac{m^2(1-\beta)}{\alpha(1-\alpha)}-p^2\beta^2\right\}}+Ap^2+B\right]. \qquad (2.66)$$

This expression, though complicated, is capable of some

simplification, but not all the integrals may be done in closed form; thus we do not attempt the simplification. The expression is included for reference and completeness. A and B are constants to be fixed by specifying $I(p^2 = m^2)$ and $(\partial I/\partial p^2)(p^2 = m^2)$; they appear for the same reason as the constant C in the previous one loop calculation. We wish now to show how to introduce the counter term into the Lagrangian to eliminate the infinite part. There will be two counter terms this time. To describe how these counter terms arise it is sensible and convenient to note that the graph of Fig. 6 that we have been calculating is the second order contribution to the inverse of the propagator $\Delta_F(p, m)$. It makes sense to include the inverse bare propagator which is the first term in the perturbation expansion for $\Delta_F^{-1}(p, m)$ and write down the first two terms in the expansion

$$\Delta_F^{-1}(p, m) = -i(p^2 - m^2) + \;\;-\!\!\bigcirc\!\!-\;\; + \cdots$$
$$= -i(p^2 - m^2) + i(B'(D, \lambda)p^2 - C'(D, \lambda)m^2) + \text{Regular part} \quad (2.67)$$

where the constants $B'(D, \lambda)$ and $C'(D, \lambda)$ will have infinite parts as $D \to 4$. If we study 2.67 we notice that the presence of $C'(D, \lambda)$ has shifted the position of the zero of $\Delta_F^{-1}(p, m)$ (i.e. the pole in $\Delta_F(p, m)$) away from $p^2 = m^2$; also the presence of $B'(D, \lambda)$ has changed the residue of the pole in $\Delta_F(p, m)$ from i to some number. These changes may be compensated for by introducing two counter terms into L just as we introduced a single counter term before. The adjusted Lagrangian is

$$L' = \tfrac{1}{2}(\partial_\mu \phi)^2 - \frac{m^2 \phi^2}{2} - \frac{\lambda \phi^4}{4!} + \frac{A\lambda \phi^4}{4!} - \frac{B(\partial_\mu \phi)^2}{2} + \frac{Bm^2 \phi^2}{2} - \frac{\delta m^2 \phi^2}{2}$$
$$= L + \frac{A\lambda \phi^4}{4!} - \frac{B(\partial_\mu \phi)^2}{2} + \frac{Bm^2 \phi^2}{2} - \frac{B \delta m^2 \phi^2}{2}. \quad (2.68)$$

The way these constants perform their function is simple; δm^2 is the infinite mass counter term which adjusts for the shift in the position of the pole and B cancels the infinite part which has arisen in the residue. The new position of the pole is $m^2 + \delta m^2$, this of course is infinite but this does not worry us yet. The infinite constants $B'(D, \lambda)$ and $C'(D, \lambda)$ are easily calculated from 2.65; the final result is trivially, and more usefully, written in terms of B and m^2, and is

$$B = \frac{\lambda^2}{12(16\pi^2)^2(D-4)}$$

$$\delta m^2 = -\frac{5\lambda^2 m^2}{12(16\pi^2)^2(D-4)} - \frac{2\lambda^2 m^2}{(16\pi^2)^2(D-4)^2}. \qquad (2.69)$$

We draw the reader's attention to the double pole in 2.69, the existence of which we had already been made aware earlier in the calculations. It is now appropriate to discuss the qualitative nature of the result. Firstly, to summarize what we have achieved thus far: the four-point function and the two-point function (inverse propagator), have been calculated to the lowest non-trivial order in perturbation theory. It has proved possible to cancel their infinite parts by the introduction of counter terms into the original Lagrangian. The finite parts are then clearly isolated. It should now be clear that this procedure can be repeated for all the higher orders of perturbation theory. For example, if the four- and two-point functions were expanded still further to include the graphs shown below, then if the counter terms are extended to higher orders in λ the same counter terms will be able to cancel the divergent parts of the graphs. Note that the two sources of divergence may occur in a mixed form as shown in the second graph of Fig. 9(a). The above prompts us to write down the general expressions for

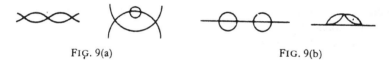

FIG. 9(a)   FIG. 9(b)

the quantities C, B and δm^2 which make up the counter terms. These are

$$C = \sum_{n=1}^{\infty} \sum_{m=n}^{\infty} C_{nm} \frac{\lambda^m}{(D-4)^n}$$

$$B = \sum_{n=1}^{\infty} \sum_{m=n}^{\infty} b_{nm} \frac{\lambda^m}{(D-4)^n} \qquad (2.70)$$

$$\delta m^2 = \sum_{n=1}^{\infty} \sum_{m=n}^{\infty} m^2 \frac{d_{mn} \lambda^m}{(D-4)^n}$$

where we have allowed for, as we clearly must, multiple poles in D of arbitrarily high order.

Renormalization in General

Having dealt with the problem of removing infinities from the two and four point functions it is fair to ask if a similar technique with presumably more counter terms is required for the n-point functions for values of $n = 6, 8, 10, \ldots$. Happily when a theory is renormalizable this is not so, and while the higher n-point functions are infinite, no new counter terms are needed to remove these infinities; this is because they are recurrences of infinities from the two sources that we have already analysed. We shall not prove this statement but shall make it quite clear how it comes about. A very useful number associated with a Feynman graph is what is called P the degree of divergence of the graph. In the language of the cut-off method a graph whose degree of divergence[†] is P will diverge as Λ^P. Because each Feynman graph, with a fixed number of external legs n say, has a fixed dimension, $4-n$, the growth of the graph as a function of Λ is determined by this and so we define

$$P = 4a - 2b$$
$$= 4 - n \qquad (2.71)$$

where a is the number of loop integrations and b is the number of internal lines. The fact that $4a - 2b = 4 - n$ is easily proved; one first uses an inductive argument to show that $4a - 2b$ only depends on the number of external legs of a Feynman graph and not on the potentially very complex internal structure; the result then follows.

Now for some examples

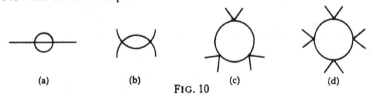

FIG. 10

We can readily make use of 2.71 to show that the degrees of divergence of Fig. 10(a) ... (d) are given by $P = 2, 0, -2, -4$ respectively. Graphs (a) and (b) we have already analysed, and have been found to be quadratically and logarithmically divergent respectively. Moreover a

---

[†] We remind the reader that when $P = 0$ the graph will in general diverge logarithmically.

simple power-counting exercise shows us that the Feynman integrals for graphs (c) and (d) are finite. It would indeed be fortunate if the property that this result suggests were true in general; i.e. if P were negative that the graph must be finite. This illusory property is quickly dispelled by counter example. Consider the graphs shown below

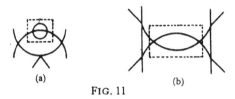

FIG. 11

these graphs have $P = -2$, -4 respectively. They are not, however, finite. To see this one examines the subgraphs enclosed by the dotted lines which are, as we already know, divergent; hence the whole Feynman graph is divergent. This situation is ameliorated considerably by the fact that these subgraphs can be made finite by using the counter terms already introduced into the theory. It also must be true that if, in any Feynman graph, there exists a subgraph whose degree of divergence is non-negative (i.e. zero or positive) then the Feynman graph is divergent. The converse of this is the germ of the general theorem that we are seeking about the convergence properties of an arbitrary Feynman graph. This general theorem, which we state without a proof is due to Weinberg,[1] and says:

Theorem. *The general Feynman graph converges if the degree of divergence P together with the degree of divergence of all subgraphs is negative.*

The proof itself is difficult but the idea that such a theorem could be true is adequately illustrated by the preceding examples. To recapitulate, although all scattering amplitudes calculated with Feynman graphs may have infinite parts, there are, (in $\lambda\phi^4/4!$ theory), only two genuinely distinct sources of divergence, namely those that we have already analysed, the two-point function and the four-point function. These divergences are removed by use of the counter terms given in eq. 2.70.

The time has now come to relate the counter terms to the renormalization constants of Chapter 1. We recall that if we use renormalized fields, masses and coupling constant we can transfer all the content of renormalization into the infinite renormalization constants. For $\lambda\phi^4/4!$ theory this took the form

$$L_0 = Z_3\tfrac{1}{2}\{(\partial_\mu\phi_R)^2 - m_R^2\phi_R^2\} - Z_1\frac{\lambda_R}{4!}\phi_R^4 + Z_3\frac{\delta m^2\phi_R^2}{2} \qquad (2.72)$$

where the relations between the renormalized and unrenormalized quantities are

$$\phi_0 = Z_3^{1/2}\phi_R$$
$$\lambda_0 = \frac{Z_1}{Z_3^2}\lambda_R \qquad (2.73)$$
$$m_R^2 = m_0^2 + \delta m^2$$

and the subscripts 0 and R denote unrenormalized and renormalized quantities respectively. We have purposely avoided using the subscripts 0 and R on the quantities used in the previous work of this chapter. This is because we were involved only with elucidating the infinities. But we see that if we wish we can absorb all the counter terms into definitions involving renormalized quantities. To do this we rewrite L_0 as

$$L_0 = \tfrac{1}{2}\{(\partial_\mu\phi_R)^2 - m_R^2\phi_R^2\} - \frac{\lambda_R}{4!}\phi^4 - (Z_1-1)\frac{\lambda_R}{4!}\phi^4$$
$$+ \frac{(Z_3-1)}{2}\{(\partial_\mu\phi_R)^2 - m_R^2\phi_R^2\} + Z_3\frac{\delta m^2}{2}\phi_R^2$$
$$= L_R - (Z_1-1)\frac{\lambda_R}{4!}\phi^4 + (Z_3-1)\tfrac{1}{2}\{(\partial_\mu\phi_R)^2 - m_R^2\phi_R^2\} + Z_3\,\delta m^2\frac{\phi_R^2}{2}$$
$$(2.74)$$

with the previous definition for L_R. Now the Lagrangian L_R written in terms of the physical renormalized quantities gives finite results. Hence L_R is L', the finite Lagrangian that we constructed using counter terms, and L_0 is the quantity that we write as L. The subtraction constants A and B then have the most convenient definitions

$$A = (Z_1-1)$$
$$B = (Z_3-1) \qquad (2.75)$$

comparing 2.74 with 2.68, (the m^2 appearing in 2.68 is, of course, m_0^2).

Thus we see that we may readily relate m_R^2 to m_0^2, or find exact expressions for Z_1 and Z_3 by using the expressions for A, B and δm^2 already obtained. The physical point to be emphasized is that all infinities reside in the renormalization constants Z_1, Z_3 and in δm^2; but these quantities relate the bare quantities whose magnitude is indeterminate to the renormalized ones. We then resolve this indeterminacy by identifying the renormalized quantities m_R^2, λ_R ... etc. to be the physical quantities as measured in some well defined experimental situation. Further these physical renormalized quantities can be used to construct a Lagrangian L_R in the use of which all infinities cancel and only finite physical results remain. This is the content of renormalization theory.

A Counter Example

Before ending this chapter we wish to indulge in a minor digression which would have disrupted our discussion if we had introduced it earlier. We wish to dispel the idea that the finite parts of Feynman integrals are obtained in the dimensional method by simply discarding poles and multiple poles at $D = 4$. To see this consider the graph shown in Fig. 12. Apart from a numerical factor proportional to λ this

FIG. 12

graph should simply be the square of the single loop graph of Fig. 4.†
Let us call the finite part of the graph of Fig. 4 $I(D)$ and that of Fig. 12 $J(D)$ then we should have

$$J(D = 4) = C\lambda I^2(D = 4) \qquad (2.76)$$

for some constant C. To see what in fact happens we write

$$\hat{I}(D) = I_1(D-4) + I + \frac{I_2}{(D-4)} \qquad (2.77)$$

$\hat{I}(D)$ is the general expression for the graph whose finite part in four dimensions is I. Then $J(D = 4)$ should be given by $C\lambda(\hat{I}^2(4))_F$ where

---

† This is because we believe unitarity to hold order by order in power of λ.

$(\hat{I}^2(4))_F$ is the finite part of $\hat{I}^2(D)$ at $D = 4$. However from 2.77 we compute

$$(\hat{I}^2(4))_F = I^2 + 2I_1 I_2 \neq I^2. \tag{2.78}$$

This result violates unitarity if true, and the way round it is to introduce some mechanism which makes graphs finite step by step, rather than making everything finite at the end. The most attractive way of doing this is to introduce a parameter S and replace the counter term $A(\lambda\phi^4/4!)$ by $SA(\lambda\phi^4/4!)$. If we now work out finite parts order by order in S, and set $S = 1$ at the end then this automatically solves our problem. Clearly the other counter terms must be treated similarly. It may be disappointing that finite parts are not obtained by discarding the poles in the unrenormalized graphs, but at least the fact that it is not so is not a disaster.

This brings us to the close of this chapter. In the next chapter we shall discuss the renormalization of QED and later on a general criterion for deciding on the renormalizability of a Lagrangian field theory.

References

1. S. Weinberg, *Phys. Rev.*, **118**, 838 (1960).
2. S. L. Adler, *Phys. Rev.*, D5, 3021 (1972); The perturbation series of $\lambda\phi^4/4!$ theory diverges in 2-dimensional space-time, *see* A. Jaffe, *Comm. Math. Phys.*, **1**, 127 (1965).
3. R. J. Eden, P. V. Landshoff, D. I. Olive and J. C. Polkinghorne, 'The Analytic S-Matrix', Cambridge University Press (1966).
4. G. 't Hooft and M. Veltman, *Nucl. Phys.*, B44, 189 (1972).
5. G. 't Hooft, *Nucl. Phys.*, B33, 173 (1971), B35, 167 (1971).

THREE

Dimensional Regularization of Quantum Electrodynamics

Quantum Electrodynamics: Lagrangian and Feynman Rules

In this chapter we examine the renormalization properties of Quantum Electrodynamics, QED. QED is a more complicated theory and its divergences reflect these complications. The two main sources of complication are the presence of Fermions, $\psi(x)$, as well as bosons, $A_\mu(x)$, and the property of local gauge invariance. There is also another new feature deriving from the masslessness of the photon $A_\mu(x)$: this is that there are infrared divergences in some Green's functions. These are entirely separate from the divergences that are encountered in renormalization. They can be dealt with and the resultant S-matrix elements are perfectly finite. The subject of the infrared divergences of QED and the gauge properties of QED will be covered in more detail in Chapters 4 and 5.

The Lagrangian for QED is given below

$$L = -\tfrac{1}{4}F_{\mu\nu}F^{\mu\nu} - \tfrac{1}{2}(\partial_\lambda A^\lambda)^2 + \bar{\psi}(i\gamma_\mu \partial^\mu - m - e\gamma_\mu A^\mu)\psi. \qquad (3.1)$$

This gives rise to the Feynman rules given in Chapter 1. We repeat these for convenience.

internal electron line	——→—— p	$\dfrac{i}{\not{p} - m + i\varepsilon}$
internal photon line	μ ⌇⌇⌇ ν p	$-i\dfrac{g_{\mu\nu}}{p^2 + i\varepsilon}$
vertex		$-ie\gamma_\mu$
closed Fermion loop		— Trace
loop momentum k		$\displaystyle\int \dfrac{dk}{(2\pi)^4}$
external electron line		$\bar{u}(p,s)$, outgoing $u(p,s)$, incoming
external photon line		ε_μ

where ε_μ is the polarization vector for the photon and must be transverse for real photons and $F_{\mu\nu}$ is the usual electromagnetic tensor $\partial_\mu A_\nu - \partial_\nu A_\mu$. It will not have escaped the notice of the well informed reader that the term $-\tfrac{1}{2}(\partial_\lambda A^\lambda(x))^2$ in the Lagrangian is one which usually has no classical counterpart. It is called the gauge fixing term and its properties will be discussed in detail later. For the moment we simply accept its presence but note that it does not introduce any new vertices into the Feynman graphs but merely affects the photon propagator.

Renormalization of QED

It is now the time to ask if the Lagrangian 3.1 gives rise to a renormalizable theory of quantum electrodynamics. Happily, as we shall see, the answer is yes. The method of dimensional regularization is also applicable and is of great use in maintaining gauge invariance; it is also in general quicker and more elegant than other methods such as one which uses ultraviolet cut-offs. First of all we provide some examples of divergent graphs; these are shown below.

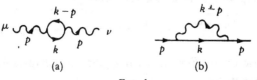

FIG. 1

An examination of the Feynman integrals for the graphs of Fig. 1 quickly reveals them both to be divergent and, of course, there are many other divergent graphs. However with the benefit of the discussion and treatment of $\lambda\phi^4/4!$ theory in Chapter 2 behind us, we have a guide as to how to develop our analysis of these divergences. What we learned in the last chapter was that there were only a finite number of independent sources of divergences, and, though these could be combined in an infinite variety of ways, the same method for dealing with them was always used. The most informative number about a graph, the degree of divergence, can tell us whether it has a possibility of being finite. Further, if the degree of divergence of all the subgraphs of a Feynman graph is also known then we can actually answer the question of finiteness. The obvious inference that one draws from the lesson of the last chapter is that one should define a degree of divergence P for QED as was done for $\lambda\phi^4/4!$ theory. We shall do this straight away so that the degree of divergence P of a graph in QED is the power of the ultraviolet cut-off Λ with which the graph diverges as Λ becomes infinite. As in $\lambda\phi^4/4!$ theory this may be related to the internal structure of the graph in terms of numbers of internal lines and numbers of loops. The dimension of the graph in powers of mass is also equal to P and if we realize this then it is possible to see that

$$P = 4j - 2a - b \qquad (3.2)$$

where j is the number of loop integrations, a is the number of internal photon lines, and b is the number of internal Fermion lines. In addition to this, because P is the dimension of the graph, and because S-matrix elements have a dimension which is fixed when the number of scattered particles is fixed; then P only depends on the number of external lines of the graph. This gives us the formula

$$P = 4 - m - 3n/2 \qquad (3.3)$$

where m is the number of external photon lines and n is the number of external Fermion lines. The presence of $3n/2$ in 3.3 should not cause any trouble since a little thought will show that the number of external Fermion lines in a graph is necessarily an even number. The task now is to systematically write down graphs with increasing numbers and configurations of external lines until one has found all those for which $P \geq 0$. This is a simple matter to do and we show these in Fig. 2.

The box represents a complete sum of Feynman graphs and the degree of divergence P is written inside each box. For all other configurations of external lines P is negative. Perusal of Fig. 2 would suggest that there are five independent sources of divergence to contend with in QED; fortunately this is not so. The number may be quickly reduced to four by noticing that Fig. 2(c) is an interaction of an odd number (three) of photons and this is forbidden by the well-known Furry's

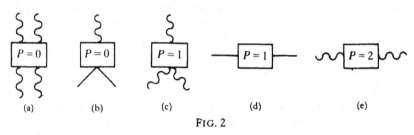

FIG. 2

theorem (proved in reference (1)) or by recalling that the photon has odd charge conjugation.† Thus Fig. 2(c) is identically zero. Next on the list for removal is Fig. 2(a); this amplitude must satisfy gauge invariance requirements which are that the tensor structure of the external lines is constrained to be of a certain form. This form is such that if $M_{\lambda\mu\nu\sigma}(p, q, r, s)$ is the amplitude of Fig. 2(a) where p, q, r and s are the external momenta and λ, μ, ν, σ the respective tensorial indices then $M_{\lambda\mu\nu\sigma}(p, q, r, s)$ satisfies‡

$$p^{\lambda}M_{\lambda\mu\nu\sigma} = q^{\mu}M_{\lambda\mu\nu\sigma} = r^{\nu}M_{\lambda\mu\nu\sigma} = s^{\sigma}M_{\lambda\mu\nu\sigma} = 0 \qquad (3.4)$$

and other related identities. To do this the amplitude must decompose into sums of quantities which are products of external momenta with tensorial indices times a Lorentz scalar. This means that the resultant scalar quantities have $P < 0$ (in fact $P = -4$) and so are convergent. The list in Fig. 2 now consists of the three graphs in Figs 2(b), (d) and

---

† Under charge conjugation C the photon obeys the equation $CA_\mu x(C)^{-1} = -A_\mu(x)$, this together with the equality $C|0\rangle = |0\rangle$ means that

$$\langle 0|A_{\mu_1}A_{\mu_2}\ldots A_{\mu_{2n+1}}|0\rangle = \langle 0|CA_{\mu_1}C^{-1}CA_{\mu_2}\ldots CA_{\mu_{2n+1}}C^{-1}|0\rangle$$
$$= (-1)^{2n+1}\langle 0|A_{\mu_1}\ldots A_{\mu_{2n+1}}|0\rangle$$

thus $\langle 0|A_{\mu_1}\ldots A_{\mu_{2n+1}}|0\rangle$ is zero.

‡ The reader is urged to verify the properties of $M_{\lambda\mu\nu\sigma}$ by computing the trace in the contribution to Fig. 2(a) which consists of one closed loop with four external lines. Equation 3.4 is proved in Chapter 4.

(e). For these graphs there is no fortuitous circumstance which makes them convergent, but nevertheless they represent only two and not three independent divergences, contrary to what one might at first sight expect. The reason is that the vertex (Fig. 2(b) and the Fermion propagator Fig. 2(d) are related by the Ward† identity, and if the infinite part of one is calculated, the infinite part of the other may also be deduced by using the Ward identity. Thus in summary there are only two distinct sources of divergence in QED and these may be taken to be those of Fig. 2(e) and (b), or those of Fig. 2(e) and (d), whichever one prefers. A point worth mentioning when discussing the nature of the divergences in QED is that there is, as there was in the last chapter, an overlapping divergence. This occurs in its simplest form in the graph shown in Fig. 3. Nothing essentially new has been added to the argument of the last chapter to eliminate this divergence.

FIG. 3

The Ward Identity

This is a topic that we shall return to in the next chapter, where we shall use the functional methods of Chapter 1 to provide a more formal and wide-ranging discussion, which is suitable for expansion as a general method for treating theories with a local gauge invariance. Here, however, we shall use the more familiar diagrammatic analysis, which cannot, in any proper treatment of the Ward identity, be omitted. It was from a direct study of the Feynman diagrams that the matter was first understood and it still remains the most valuable intuitive method for discussing the Ward identity. This being so there is in fact a formulation of gauge theories such as QED and non-Abelian Yang–Mills' theories in which Feynman diagrams are the basis from which all else follows. The difficult and subtle treatment of gauge invariance for Yang–Mills' theories can be rigorously formulated in terms of rules for diagrams.[2]

First of all the Ward identity is a relation which connects the inverse Fermion propagator, $S_F^{-1}(p)$, and the Fermion-photon vertex, $\Gamma_\mu(p, q, p+q)$. To lowest order the bare versions of these quantities

† See the next section in this chapter.

are given by (0 denoting bare)

$$S_F^{-1_0}(p) = \slashed{p} - m, \qquad \Gamma_\mu^0(p, q, p+q) = \gamma_\mu. \qquad (3.5)$$

Simple inspection tells us that to this order we may write

$$\frac{\partial S_F^{-1_0}(p)}{\partial p_\mu} = \Gamma^0(p, q, p+q). \qquad (3.6)$$

To see whether this is an interesting fact which, though true for the bare propagator and vertex, may not be true in general, we investigate the next order in perturbation theory. We show this graphically in Fig. 4.[†]

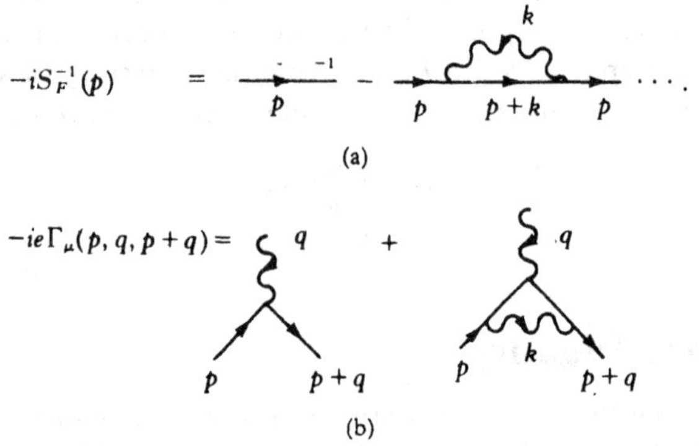

FIG. 4

Continuing our investigation we compute $\partial S_F^{-1}/\partial p_\mu$ from Fig. 4(a) and obtain

$$\frac{\partial S_F^{-1}}{\partial p_\mu} = \gamma_\mu + \frac{ie^2}{(2\pi)^4} \int dk\, u(p)\gamma_\alpha \frac{1}{(\slashed{p}+\slashed{k}-m)} \gamma_\mu \frac{1}{(\slashed{p}+\slashed{k}-m)} \gamma^\alpha \bar{u}(p). \qquad (3.7)$$

But from Fig. 4(b) we find that

$$\Gamma_\mu(p, q, p+q) = \gamma_\mu + \frac{ie^2}{(2\pi)^4} \int dk\, u(p)\gamma_\alpha \frac{1}{(\slashed{p}+\slashed{k}-m)} \gamma_\mu \frac{1}{(\slashed{p}+\slashed{k}+\slashed{q}-m)}$$
$$\cdot \gamma^\alpha \bar{u}(p). \qquad (3.8)$$

[†] The minus sign in Fig. 4(a) needs clarification. It is present because to obtain Fig. 4(a) we expand the propagator, rather than its inverse to second order. We then invert this and obtain an expression of the form $1/(1+x)$ where x is the second order one loop graph shown in Fig. 4(a), expanding this in powers of x gives to lowest order $1-x$ as given above.

DIMENSIONAL REGULARIZATION OF QUANTUM ELECTRODYNAMICS 97

We conclude immediately by comparing 3.7 and 3.8 that 3.6 is no longer true and hence cannot be true in general. It is nevertheless true that if the momentum q of the photon is zero then the similar equation

$$\frac{\partial S_F^{-1}}{\partial p_\mu}(p) = \Gamma_\mu(p, 0, p) \qquad (3.9)$$

is true, to this order at least. This is the form of the equation that we have been after, and this equation is actually true to all orders in perturbation theory. We can understand this in a simple fashion. The result hinges on realizing that when the internal Fermion lines are differentiated with respect to p_μ, one always obtains a bare vertex γ_μ sandwiched between two Fermion propagators of equal momentum; this is, of course, a vertex with a zero-momentum photon. This point having been established one repeats the argument for all internal Fermion lines in a given diagram; this gives a contribution to the zero-momentum vertex of the same order in e. In this way one may produce an inductive argument for the proof of 3.9 to all orders in the coupling constant e.

The verification of 3.9 must be considered carefully when, as will nearly always be the case, divergent graphs contribute to $S_F^{-1}(p)$ and $\Gamma_\mu(p, 0, p)$. What happens then is that S_F^{-1} and Γ_μ must first be made finite before any sense can be made of eq. 3.9. This means that we continue S_F^{-1} and Γ_μ away from $D = 4$. With the analytic continuation that we use in this book we shall find that 3.9 does indeed hold. In this context though a note of caution must be supplied. The analytic continuation that we use is not unique; this is because we have analytically continued to arbitrary complex D integrals which were only defined for integer D. For example, if we were to add to the continuation of $\Gamma_\mu(p, 0, p)$ the quantity $\sin(\pi D)$ this would not change $\Gamma_\mu(p, 0, p)$ for integral D. It would, however, change $\Gamma_\mu(p, 0, p)$ for non-integral D and would make 3.9 no longer true. So as regards the Ward identity 3.9 we are faced with somewhat of a dilemma. Equation 3.9 diverges for $D = 4$; however if we analytically continue away from $D = 4$ to make everything finite, then 3.9 may or may not hold due to the lack of uniqueness of this procedure. What we actually do is to require 3.9 to hold even for D non-integral; this results with the method of continuation described in Chapter 2. This continuation seems to be the most natural one to use, but, of course, demanding gauge invariance to be satisfied for the finite renormalized

Feynman amplitudes is an act of faith, which has a basis in the formal properties of gauge invariance, such as those derived in Chapter 4, which hold if one ignores divergence difficulties. Difficulties of this kind are inevitable if one begins with a divergent theory and tries to construct from it a finite theory. When we claim elsewhere in the text to have satisfied the requirements of gauge invariance we still have this proviso in mind. Equation 3.9, although very useful, is not the most general form of the identity that there is; 3.9 relates the derivative of $S_F^{-1}(p)$ to $\Gamma_\mu(p, 0, p)$, it is possible to derive an equation relating $S_F^{-1}(p)$ itself to $\Gamma_\mu(p, 0, p')$. The more general form that we seek we shall state here and prove in the next chapter. The Ward identity is

$$S_F^{-1}(p+q) - S_F^{-1}(p) = q^\mu \Gamma_\mu(p, q, p+q). \qquad (3.10)$$

It is easy, as it was above, to check that 3.10 is true for the bare propagator and bare vertex. The fact that it is trivially true for the bare propagator and bare vertex provides the clue, as it did above, to the method for establishing the result in general. One uses the diagrammatic inductive analysis on each internal Fermion line in contributions to $S_F^{-1}(p)$ and $S_F^{-1}(p+q)$ and relates them to contributions to $\Gamma_\mu(p, q, p+q)$. We have mentioned that 3.9 only relates the derivative of S_F^{-1} and Γ_μ. As one can see, 3.10 relates a form of S_F^{-1} to Γ_μ with no derivative. This is a great advantage as we can by judicious choice of momentum turn eq. 3.10 into an equation which allows us to compute $S_F^{-1}(p)$ from Γ_μ. It is quite simple to do this, and the argument runs as follows: since the propagator $S_F(p)$ has a pole at the physical mass-shell point $p^2 = m^2$ then $S_F^{-1}(p)$ is zero there.† Thus we may obtain $S_F^{-1}(p)$ from 3.10 by evaluating it at the point $p^2 = m^2$ obtaining

$$S_F^{-1}(p+q) = q^\mu \Gamma_\mu(p, q, p+q), \qquad p^2 = m^2. \qquad (3.11)$$

This equation is really a matrix equation for two scalar functions A and B and we wish to display it in its uncoupled form. The inverse propagator gives rise to two scalar functions in its Dirac matrix structure (only two because all others are forbidden by the general requirement of parity invariance and Lorentz invariance); these are defined by

$$S_F^{-1}(p) = \not{p} A\left(\frac{p^2}{m^2}\right) - \operatorname{Im} B\left(\frac{p^2}{m^2}\right). \qquad (3.12)$$

† This is actually not true in all gauges because of the infrared problem; this will be analysed when the infrared problem is discussed.

Equation 3.12 and the properties of Dirac matrices used in this discussion are proved in the next section of this chapter. With the results of this section we have that

$$A = \frac{1}{4p^2} \operatorname{Tr} \{\slashed{p} S_F^{-1}(p)\}$$
$$-mB = \tfrac{1}{4} \operatorname{Tr} \{S_F^{-1}(p)\}. \tag{3.13}$$

Equations 3.11, 3.12, 3.13 provide the uncoupled form of the equations for A and B. We now include for completeness the fairly obvious remark that the limiting form of 3.10 as the momentum q_μ tends to zero implies our first form of the Ward identity, i.e. eq. 3.9. Our final equations relating A and B to Γ_μ are

$$A\left\{\frac{(p+q)^2}{m^2}\right\} = \frac{1}{4(p+q)^2} \operatorname{Tr} \{(\slashed{p}+\slashed{q}) q^\mu \Gamma_\mu(p, q, p+q)\}, \quad p^2 = m^2$$
$$-mB\left\{\frac{(p+q)^2}{m^2}\right\} = \tfrac{1}{4} \operatorname{Tr} \{q^\mu \Gamma_\mu(p, q, p+q)\}, \quad p^2 = m^2. \tag{3.14}$$

We have discussed the Ward identities in this sort of detail because a knowledge of them is invaluable for acquiring manipulative skill in evaluating complicated diagrams in QED, or in estimating their asymptotic behaviour in certain kinematical limiting situations. We also wish to use them to relate the counter terms needed for the propagator and the vertex so that we may set up the renormalization scheme for QED. For this purpose we shall need only make use of 3.9. To investigate the consequences of the Ward identity for the renormalization of QED we shall first assume the theory to be renormalizable and derive the relation that the Ward identity implies among the renormalization constants. Then we shall derive, to the lowest nontrivial order, expressions for the renormalization constants and prove, with the same level of rigour as in the last chapter, that these obey the relation implied by the Ward identity.

Recalling the discussion of renormalization constants in Chapter 1, we derived the form for $S_F(p)$ near its mass shell

$$S_F(p) \underset{p^2 \to m^2}{\to} \frac{Z_2}{\slashed{p} - m} \tag{3.15}$$

$$S_F^{-1}(p) \underset{p^2 \to m^2}{\to} Z_2^{-1} (\slashed{p} - m). \tag{3.16}$$

On the other hand when we consider the Fermion–photon vertex we introduce a counter term $A = Z_1 - 1$ (using the same notations as the last chapter) to make it finite. As we shall see below the effect of the counter term may be stated as a multiplicative modification of the vertex when evaluated at zero momentum transfer. That is to say the finite vertex obeys the equation

$$\Gamma_\mu(p,0,p) + A\Gamma_\mu(p,0,p) = \gamma_\mu \underset{p^2=m^2}{=} Z_1 \Gamma_\mu(p,0,p). \quad (3.17)$$

Inserting 3.16, 17 into 3.17 and taking the limit $p^2 \to m^2$ we discover immediately that

$$Z_2 = Z_1. \quad (3.18)$$

This is the relation that we sought and it means that we may renormalize the theory with two constants, Z_1 and Z_3 (Z_3 being the photon wave function renormalization constant), or Z_2 and Z_3, whichever we prefer. The next matter to examine is the applicability of the dimensional regularization method to QED followed by the construction and calculation of the counter terms needed for QED.

Dimensional Method for QED

The way to conduct this discussion is to compare the problem of regularizing QED diagrams with diagrams of the $\lambda\phi^4/4!$ theory. For a specific example we choose to compare two sets of diagrams

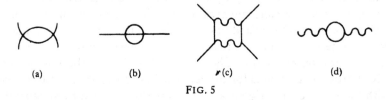

(a)　　　(b)　　　(c)　　　(d)

FIG. 5

If we compare the four-point functions that are shown in Fig. 5(a) and (c) we find that Fig. 5(c) has the complications of Dirac matrices on the internal lines and the vertices, and spinor wave functions u and $\bar{u}$ on the external lines, and there are also tensorial indices on the vertices and the photon propagators. Thus we must learn how to deal with the continuation to arbitrary D in the presence of all these quantities. If we compare Figs 5(b) and (d) the complications are the presence of tensorial indices and the trace over the Dirac matrices of the internal Fermion loop. So this demands that we learn how to continue Feynman diagrams to complex D in the presence of a trace over Dirac matrices. When stated in this way the problem that we are faced with looks quite non-trivial. It is actually less complex than this discussion would make it appear. The correct thing to begin doing is to ask if there are any features in the above that are independent of the space time dimension D. There are, and these are the structure of the anticommutation relation for the Dirac matrices and also the properties of traces of products of the Dirac matrices. The anti-commutation relation for the Dirac γ matrices is

$$\gamma_\mu \gamma_\nu + \gamma_\nu \gamma_\mu = \{\gamma_\mu, \gamma_\nu\} = 2g_{\mu\nu} I. \tag{3.19}$$

In 3.19 the numerical factor 2 has nothing to do with the dimension of space-time D, but the tensor $g_{\mu\nu}$ does. If we now turn to the trace we discover that

$$\mathrm{Tr}\,[\gamma_\mu \gamma_\nu] = \tfrac{1}{2}\,\mathrm{Tr}\,[\gamma_\mu \gamma_\nu + \gamma_\nu \gamma_\mu] = g_{\mu\nu}\,\mathrm{Tr}\,I = 4g_{\mu\nu}. \tag{3.20}$$

where we have used the property that $\mathrm{Tr}\,AB = \mathrm{Tr}\,BA$ for any finite-dimensional matrices A and B. In 3.20 the numerical factor 4 is simply the trace of the identity matrix. To deal with a general trace of a product of n matrices we can use 3.19, 3.20 to derive an induction formula

$$\mathrm{Tr}\,[\gamma_{\mu_1} \ldots \gamma_{\mu_n}] = 2g_{\mu_1\mu_2}\,\mathrm{Tr}\,[\gamma_{\mu_3} \ldots \gamma_{\mu_n}] - \mathrm{Tr}\,[\gamma_{\mu_2}\gamma_{\mu_1}\gamma_{\mu_3} \ldots \gamma_{\mu_n}] \tag{3.21}$$

where we have used the anti-commutation relation 3.19 on the product $\gamma_{\mu_1}\gamma_{\mu_2}$. We now use 3.19 again on the product $\gamma_{\mu_1}\gamma_{\mu_3}$ to derive another equation, in this way we move the matrix γ_{μ_1} successively to the right until we arrive at a trace of the form $\mathrm{Tr}\,[\gamma_{\mu_2} \ldots \gamma_{\mu_n}\gamma_{\mu_1}]$. But by the commutativity of the trace this is equal to the LHS $\mathrm{Tr}\,[\gamma_{\mu_1} \ldots \gamma_{\mu_n}]$; thus the formula reduces the trace of the product of n γ-matrices to the sum of the traces of products of $n-2$ γ-matrices. This is the desired induction step; the final trace of the product of 2 γ-matrices is given in 3.20. Throughout this argument we have not pointed out that the integer n is necessarily even. This is so, as if we recall the well-known matrix γ_5 defined by

$$\gamma_5 = i\gamma^0 \gamma^1 \gamma^2 \gamma^3, \tag{3.22}$$

then use of 3.19 enables us to verify that

$$\{\gamma_5, \gamma_\mu\} = 0, \qquad \gamma_5^2 = I. \tag{3.23}$$

Now if we consider $\mathrm{Tr}\,[\gamma_{\mu_1} \ldots \gamma_{\mu_n}]$ where n is odd we may use 3.23 to show that

$$\mathrm{Tr}\,\{\gamma_{\mu_1}\gamma_{\mu_2}\ldots\gamma_{\mu_n}\} = \mathrm{Tr}\,\{\gamma_5^2\gamma_{\mu_1}\ldots\gamma_n\} = (-1)^n \,\mathrm{Tr}\,\{\gamma_5\gamma_{\mu_1}\ldots\gamma_{\mu_n}\gamma_5\}. \tag{3.24}$$

The factor $(-1)^n$ comes from moving γ_5 to the right n times using the commutation relation in 3.23. Then if $A = \gamma_5$ and $B = \gamma_{\mu_1}\ldots\gamma_{\mu_n}\gamma_5$ because $\mathrm{Tr}\,\{AB\} = \mathrm{Tr}\,\{BA\}$ and $\gamma_5^2 = I$ we find that

$$\mathrm{Tr}\,\{\gamma_{\mu_1}\ldots\gamma_{\mu_n}\} = (-1)^n \,\mathrm{Tr}\,\{\gamma_{\mu_1}\ldots\gamma_{\mu_n}\}. \tag{3.25}$$

But if n is odd then this means that $\mathrm{Tr}\,\{\gamma_{\mu_1}\ldots\gamma_{\mu_n}\}$ is zero. From this general analysis of the trace of γ matrices we realize that the only space-time dependent object which occurs in the traces is the tensor $g_{\mu\nu}$. We simply define the properties of $g_{\mu\nu}$ in D dimensions in the natural way to be

$$g_{\mu\nu}T^\nu_{\lambda\tau\ldots} = T_{\mu\lambda\tau\ldots} \qquad g^\mu_\mu = D \tag{3.26}$$

$$g_{\mu\nu}p^\mu p^\nu = p^\mu p_\mu = p^2$$

where $T_{\mu\lambda\tau}$ is some arbitrary tensor constructed from the tensor $g_{\mu\nu}$ and momenta p_μ. These rules are the only ones that are needed to manipulate momenta p_μ and the tensor $g_{\mu\nu}$ in D dimensions; notice that the only place where D appears is in the expression for the trace of $g_{\mu\nu}$. We have now discovered how to deal with the tensors and traces that we shall meet in Feynman diagrams; it remains to discover what to do with the external spinor wave functions. It turns out that these may all be turned into traces and so our problem is solved. To see how this comes about, consider an S-matrix element in QED with m external photon lines and $2n$ external Fermion lines.

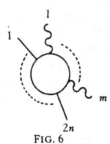

FIG. 6

If we begin with an incoming Fermion line with wave function $u(p_1)$ we can follow this line into the depths of the diagram until it emerges as an external line (it will always do this, of course) with wave function $\bar{u}(p_1')$. Doing this for all the Fermion lines provides us with an expression for the S-matrix element of the form

$$S_{fi} = \bar{u}(p_1')M_1 u(p_1)\bar{u}(p_2')M_2 u(p_2)\ldots \bar{u}(p_n')M_n(p_n) \quad (3.27)$$

where M_i, $i = 1, \ldots n$, are products of γ matrices. The rate or transition probability which is measured is

$$|S_{fi}|^2 = S_{fi}(S_{fi})^*. \quad (3.28)$$

Hence if we take the first factor in S_{fi} it contributes to 3.28 a factor (s is the spin)

$$\{\bar{u}(p_1', s')M_1 u(p_1, s)\}\{\bar{u}(p_1', s')M_1 u(p_1, s)\}^*$$
$$= \bar{u}(p_1', s')M_1 u(p_1, s)\bar{u}(p_1, s)\bar{M}_1 u(p_1', s'). \quad (3.29)$$

However, the spinor product $u(p_1, s)\bar{u}(p_1, s)$ has the well known property that it may be written as the product of the energy and spin projection operators[†]

$$u(p_1, s)\bar{u}(p_1, s) = \frac{(\not{p}_1 + m)}{2m} \frac{(1 + \gamma_5 \not{s})}{2} \quad (3.30)$$

so that 3.29 becomes

$$\bar{u}(p_1', s')M_1 \frac{(\not{p}_1 + m)}{2m} \frac{(1 + \gamma_5 \not{s})}{2} \bar{M}_1 u(p_1', s') \quad (3.31)$$

which we rewrite as

$$\sum_{\substack{\text{all 4 spinor} \\ \text{states}}} \bar{u}(p_1', s')M_1 \frac{(\not{p}_1 + m)}{2m} \frac{(1 + \gamma_5 \not{s})}{2} \bar{M}_1 \frac{(\not{p}_1' + m)}{2m} \frac{(1 + \gamma_5 \not{s}')}{2} u(p_1', s'). \quad (3.32)$$

In 3.32 we have summed over all four spinor states of positive and negative energy and spin up and spin down, but we have inserted the energy and spin projection operators to render zero the extra terms in the sum. Finally we realize that since u and $\bar{u}$ form an orthonormal

---

[†] For the definition of the polarization vector s_μ and a derivation of 3.30 see the appendix and the book by Bjorken and Drell.

basis for the four-dimensional Dirac space then if T is a matrix in Dirac space
$$\sum_{\text{all 4 states}} \bar{u}Tu = \text{Tr}\{T\}.$$

Thus 3.32 is simply

$$\text{Tr}\left\{M_1\frac{(\not{p}_1+m)}{2m}\frac{(1+\gamma_5\not{s})}{2}\bar{M}_1\frac{(\not{p}'_1+m)}{2m}\frac{(1+\gamma_5\not{s}')}{2}\right\}. \quad (3.33)$$

In this way the general transition probability $|S_{fi}|^2$ may be written as a product of traces. We note that if the electron beams, as is usual, are not polarized then the spin states will have to be summed over in the transition probability hence the spin projection operator $(1+\gamma_5\not{s})/2$ will not appear in the final traces. We would then use instead of 3.30 the relation

$$\sum_{\text{spins}} \bar{u}(p,s)u(p,s) = \frac{(\not{p}+m)}{2m}. \quad (3.34)$$

We have now learnt all that is necessary to perform computations of the Feynman diagrams of QED in D space–time dimensions. The next section is devoted to the computations of diagrams in D dimensions and the evaluation of the counter terms and thus the renormalization constants to the lowest non-trivial order in perturbation theory.

Computation in D Dimensions in QED, Ward Identity Verification and Evaluation of Renormalization Constants

We have already tracked down the sources of divergence in QED and found that there are three, only two of which are independent. We shall now present some sample calculations to let the reader see the method at work, and then calculate all three counter terms so that we may calculate the renormalization constants Z_1, Z_2 and Z_3, and verify that $Z_1 = Z_2$ so that our dimensional method automatically respects the Ward identity.

The example that we shall begin with is the vacuum polarization diagram shown below.

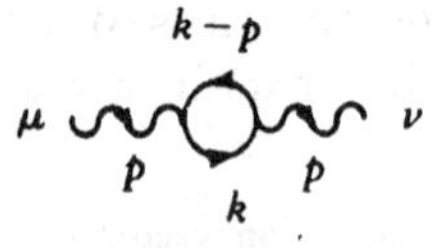

FIG. 7

Using the Feynman rules we write the contribution from Fig. 7 as

$$I_{\mu\nu} = \frac{-e^2}{(2\pi)^4} \int dk \, \text{Tr}\left\{\gamma_\mu \frac{(\slashed{k}+m)}{(k^2-m^2)} \gamma_\nu \frac{(\slashed{k}-\slashed{p}+m)}{\{(k-p)^2-m^2\}}\right\}$$

$$= \frac{-e^2}{(2\pi)^4} \int d^D k \frac{F_{\mu\nu}(k,p)}{(k^2-m^2)\{(k-p)^2-m^2\}}. \quad (3.35)$$

We use Feynman parameters to combine the denominators in the usual way and shift the variable so that

$$I_{\mu\nu} = \frac{-e^2}{(2\pi)^4} \int d^D k \int_0^1 d\alpha \frac{F_{\mu\nu}\{k+(1-\alpha)p,p\}}{[k^2+\alpha(1-\alpha)p^2-m^2]^2}. \quad (3.36)$$

To evaluate $F_{\mu\nu}$ we need an expression for the trace of the product of 4 γ-matrices. This is, using the method of the last section, easily found to be

$$\text{Tr}\{\gamma_\mu \gamma_\alpha \gamma_\nu \gamma_\beta\} = 4\{g_{\mu\alpha}g_{\nu\beta} + g_{\mu\beta}g_{\nu\alpha} - g_{\mu\nu}g_{\alpha\beta}\}$$

$$= 4S_{\mu\nu\alpha\beta}, \quad (3.37)$$

and it is readily verified that

$$F_{\mu\nu}\{k+(1-\alpha)p,p\} = 4S_{\mu\nu\alpha\beta}\{k+(1-\alpha)p\}^\alpha(k-\alpha p)^\beta + 4m^2 g_{\mu\nu}. \quad (3.38)$$

But in the integral 3.36 because the variable k has undergone a shift of origin it is only the part of $F_{\mu\nu}$ that is even in k that will contribute to $I_{\mu\nu}$. This means that we only have to evaluate the D-dimensional integrals

$$I^1_{\mu\nu} = -\frac{8e^2}{(2\pi)^4} \int dk \, d\alpha \frac{k_\mu k_\nu}{[k^2+\alpha(1-\alpha)p^2-m^2]^2}$$

$$I^2_{\mu\nu} = \frac{8e^2}{(2\pi)^4} \int dk \, d\alpha \frac{\alpha(1-\alpha)p_\mu p_\nu}{[k^2+\alpha(1-\alpha)p^2-m^2]^2}$$

$$I^3_{\mu\nu} = \frac{4e^2}{(2\pi)^4} \int dk \, d\alpha \frac{[k^2-\alpha(1-\alpha)p^2]g_{\mu\nu}}{[k_2+\alpha(1-\alpha)p^2-m^2]^2} \quad (3.39)$$

$$I^4_{\mu\nu} = -\frac{4e^2}{(2\pi)^4} m^2 \int dk \frac{d\alpha}{[k^2+\alpha(1-\alpha)p^2-m^2]^2}$$

with

$$I_{\mu\nu} = I^1_{\mu\nu} + I^2_{\mu\nu} + I^3_{\mu\nu} + I^4_{\mu\nu}. \quad (3.40)$$

We learned how to evaluate D-dimensional integrals in the last chapter and the only new formula we need is one which enables us to unravel the tensor complication in $I^1_{\mu\nu}$. The formula needed is

$$\int d^D k \frac{k_\mu k_\nu}{[k^2+a^2]^\alpha} = \frac{i\pi^{D/2}}{\Gamma(\alpha)(a^2)^{\alpha-(D/2)-1}} \frac{\Gamma(\alpha-(D/2)-1)g_{\mu\nu}}{2}. \quad (3.41)$$

This is simply derived as a D-dimensional analogue of the ordinary four-dimensional symmetric integration version.[3] With this knowledge we find for the integrals the results

$$I^1_{\mu\nu} = -\frac{4e^2}{(2\pi)^4} i\pi^{D/2} \Gamma\left(1-\frac{D}{2}\right) g_{\mu\nu} \int_0^1 \frac{d\alpha}{[\alpha(1-\alpha)p^2-m^2]^{1-(D/2)}}$$

$$I^2_{\mu\nu} = \frac{8e^2}{(2\pi)^4} i\pi^{D/2} \Gamma\left(2-\frac{D}{2}\right) p_\mu p_\nu \int_0^1 d\alpha \frac{\alpha(1-\alpha)}{[\alpha(1-\alpha)p^2-m^2]^{2-(D/2)}}$$

$$I^3_{\mu\nu} = \frac{4e^2}{(2\pi)^4} i\pi^{D/2} \Gamma\left(1-\frac{D}{2}\right) g_{\mu\nu} \int_0^1 \frac{d\alpha}{[\alpha(1-\alpha)p^2-m^2]^{1-(D/2)}} \quad (3.42)$$

$$+ \frac{4e^2}{(2\pi)^4} i\pi^{D/2} \Gamma\left(2-\frac{D}{2}\right) g_{\mu\nu} \int_0^1 d\alpha \frac{[m^2-2\alpha(1-\alpha)p^2]}{[\alpha(1-\alpha)p^2-m^2]^{2-(D/2)}}$$

$$I^4_{\mu\nu} = -\frac{4e^2}{(2\pi)^4} i\pi^{D/2} \Gamma\left(2-\frac{D}{2}\right) g_{\mu\nu} \int_0^1 d\alpha \frac{m^2}{[\alpha(1-\alpha)p^2-m^2]^{2-(D/2)}}.$$

Having separated the contributions to $I_{\mu\nu}$ in this form we discover that the quadratic divergence that we would be led to expect from dimensional analysis is present in $I^1_{\mu\nu}$ and $I^3_{\mu\nu}$. But these terms differ only in their sign so that the quadratic divergence is actually absent from $I_{\mu\nu}$ itself. Collecting coefficients of $p_\mu p_\nu$ and $g_{\mu\nu}$ we can write $I_{\mu\nu}$ as

$$I_{\mu\nu} = (p_\mu p_\nu - p^2 g_{\mu\nu}) \frac{8ie^2}{(2\pi)^4} \pi^{D/2} \Gamma\left(2-\frac{D}{2}\right)$$

$$\cdot \int_0^1 d\alpha \frac{\alpha(1-\alpha)}{[\alpha(1-\alpha)p^2-m^2]^{2-(D/2)}}. \quad (3.43)$$

The expression 3.43 for $I_{\mu\nu}$ has two advantageous properties: it has only a single logarithmic divergence, the quadratic one having disappeared, and it satisfies the condition $p^\mu I_{\mu\nu} = I_{\mu\nu} p^\nu = 0$. This latter condition is simply the requirement of gauge invariance for the electromagnetic current and it is automatically satisfied by this dimensional method. It is not as easy to ensure this requirement of gauge invariance when using the cut-off method, so that one regards the above as a very good property of the dimensional method.

To regularize $I_{\mu\nu}$ we need to separate the finite and infinite parts by the method already established. The infinite part is found to be

$$-(p_\mu p_\nu p^2 - g_{\mu\nu}) \frac{8ie^2\pi^2}{6(2\pi)^4} \frac{1}{(2-(D/2))} \quad (3.44)$$

and the finite part is found to be

$$(p_\mu p_\nu - p^2 g_{\mu\nu}) \frac{8ie^2\pi^2}{(2\pi)^4} \left[\text{const.} + \int_0^1 d\alpha\, \alpha(1-\alpha) \ln\{\alpha(1-\alpha)p^2 - m^2\} \right] \quad (3.45)$$

where the presence of the arbitrary constant was explained in the last chapter. To deal first with the infinite part, we wish to remove the divergence by adding to the Lagrangian L of eq. 3.1 a counter term. The modified form of the Lagrangian is easily found to be

$$L - A \frac{F_{\mu\nu} F^{\mu\nu}}{(-4)} \quad (3.46)$$

where

$$A = \frac{e^2}{6\pi^2(D-4)}. \quad (3.47)$$

The finite part is readily evaluated and found to be

$$(p_\mu p_\nu - p^2 g_{\mu\nu}) \frac{ie^2}{2\pi^2}$$

$$\cdot \left[C + \left(1 - \frac{2m^2}{p^2}\right)(\tfrac{1}{4} - m^2/p^2)^{1/2} \ln \left\{ \frac{[\tfrac{1}{2} + (\tfrac{1}{4} - m^2/p^2)^{1/2}]}{[\tfrac{1}{2} - (\tfrac{1}{4} - m^2/p^2)^{1/2}]} \right\} \right] \quad (3.48)$$

with C a constant. We note in 3.48 the presence of the square-root branch cut at $p^2 = 4m^2$ as required by unitarity. We also draw attention to the fact that although the factor $(1 - 2m^2/p^2)$ may look like the presence of a pole at $p^2 = 0$ this is not so. If the logarithm is expanded carefully the expression is finite at $p^2 = 0$. If this were not so then Coulomb's law would be modified. This is so because a change in 3.48 for small p^2 will affect the behaviour of the Fourier transform for large

x^2; and the behaviour of the Fourier transform for large x^2 is constrained by the requirement that Coulomb's law be valid. This is the work that is necessary to deal with the first source of divergence. The next divergence that we shall deal with is shown, to lowest non-trivial order, below.

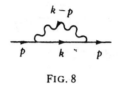

FIG. 8

The Feynman integral is readily written down and is

$$-\frac{e^2}{(2\pi)^4}u(p)\int dk\,\frac{\gamma_\mu(\slashed{k}+m)\gamma^\mu}{(k^2-m^2)\{(k-p)^2\}}\bar{u}(p) = u(p)M\bar{u}(p). \tag{3.49}$$

The Dirac matrix M may be written in the form $\slashed{k}A + mB$ for some A and B. We shall shortly find A and B, but before doing so it is convenient to have a short digression on the properties of matrices such as M, as a general method is needed if one is to be able to deal with matrices more complicated than M.

The 16 Dirac matrices are all linearly independent. This may be proved from the definitions given in the appendix. Alternatively a proof can be found in the books cited in the appendix. This being so we may choose them as a basis for the space of all 4×4 matrices since this has dimension $4^2 = 16$. Then if T is a general 4×4 matrix, it may be written as a linear combination of these 16 basis matrices. We wish to conform to convention in notation and introduce the symbol Γ_A, $A = 1, \ldots 5$, where

$$\begin{aligned}
&\Gamma_1 = I, \qquad \Gamma_2 = \gamma_5, \\
&\Gamma_3 = \gamma_\mu, \qquad \mu = 0, \ldots 3, \\
&\Gamma_4 = \gamma_5\gamma_\mu, \qquad \mu = 0, \ldots 3 \\
&\Gamma_5 = \sigma_{\mu\nu}, \qquad \mu, \nu = 0, \ldots 3, \quad \mu \neq \nu.
\end{aligned} \tag{3.50}$$

This sequence is often referred to as scalar, pseudoscalar, vector, axial vector, tensor, or SPVAT because of the obvious space–time and tensorial properties of the Γ_A. We can now express T as

$$T = \sum_{A=1}^{5} \alpha_A \Gamma^A \tag{3.51}$$

where α_A are some 16 scalar parameters to be determined. To find a formula for the α_A we refer to a very useful property of the Γ_A which is that

$$\text{Tr}\{\Gamma_A \Gamma^{A'}\} = 4\delta_{AA'}. \tag{3.52}$$

To prove 3.52, it is simply necessary to note that $\text{Tr}\,\Gamma_A = 0$ ($\Gamma_A \neq I$) and use 3.25 for the case when n is odd; 3.52 then follows. With the use of 3.52 it is seen immediately that

$$\tfrac{1}{4}\text{Tr}\{T\Gamma_A\} = \alpha_A \tag{3.53}$$

so that

$$T = \tfrac{1}{4} \sum_{A=1}^{5} \text{Tr}\{T\Gamma_A\}\Gamma^A. \tag{3.54}$$

When T is the matrix M defined in 3.49 we find that

$$M = \cancel{k}A + mB \tag{3.55}$$

with

$$\begin{aligned}\cancel{k}A &= (2-D)\frac{(-e^2)}{(2\pi)^4} \int dk \frac{\cancel{k}}{(k^2-m^2)(k-p)^2} \\ B &= D\frac{(-e^2)}{(2\pi)^4} \int \frac{dk}{(k^2-m^2)(k-p)^2}.\end{aligned} \tag{3.56}$$

The integrals in 3.56 are readily evaluated with the aid of Feynman parameters; for the infinite parts we find

$$\frac{ie^2\cancel{p}}{8\pi^2(D-4)}, \qquad -\frac{ie^2 m}{2\pi^2(D-4)} \tag{3.57}$$

for $\cancel{k}A$ and mB respectively. For the finite parts† we obtain the

---

† A general point worth noting in calculations of finite parts is that in D dimensions the coupling constant e^2 has the dimension $(m)^{(D-4)/(-2)}$; thus if we choose to work with the coupling constant $e^2(m)^{(D-4)/(-2)}$ the dimensions of quantities are automatically conserved (cf. also remarks in Chapter 2).

integrals

$$\frac{ie^2}{8\pi^2}\int_0^1 d\alpha\,(1-\alpha)\ln\left[\frac{m^2\alpha^2}{m^2\alpha-p^2\alpha(1-\alpha)}\right]\not{p}+\not{p}\,\text{const.}$$

$$-\frac{ie^2}{4\pi^2}\int_0^1 d\alpha\,m\,\ln\left[\frac{m^2\alpha^2}{m^2\alpha-p^2\alpha(1-\alpha)}\right]+\text{const.} \quad (3.58)$$

for $\not{p}A$ and mB respectively. These integrals may be readily enough done in closed form, but we do not perform them now because although the results are finite as they should be there is an additional infinity present in the electron propagator which we have not yet uncovered. This infinity is an infrared rather than an ultraviolet one, and can only be uncovered when one puts together the counter terms necessary to cancel the infinite parts with the finite parts. This we now discuss. Firstly, to the order in e^2 to which we have been working, the inverse propagator of the electron is

$$-i(\not{p}-m)+i\Sigma \quad (3.59)$$

where $i\Sigma = M$, Σ is called the self-energy of the electron. The presence of infinities in Σ has been analysed in the last chapter; they are absorbed into two infinite constants δm and Z_2 (the suffix 2 is conventional for the electron). The properties of these constants are that the physical mass of the electron is $m+\delta m$, and that Z_2 is the residue of the propagator at the new location of the pole $m+\delta m$. We wish to follow the details of this process. To this end we write

$$\Sigma = \delta m + (\not{p}-m)(1-Z_2^{-1}+C(p^2)) \quad (3.60)$$

where $C(p^2)$ is a function of p^2 which is finite. C has an important property which enables us to fix its definition from 3.60 and the integrals given in 3.58. To see what this property is we place Σ into the expression for the *inverse* propagator and obtain the expression for the *propagator*

$$\frac{i}{\not{p}-m-\Sigma}=\frac{i}{\not{p}-m-\delta m-(\not{p}-m)(1-Z_2^{-1}+C(p^2))}. \quad (3.61)$$

This we rewrite as

$$\frac{iZ_2}{-Z_2\delta m+(1-Z_2 C(p^2))(\not{p}-m)}. \quad (3.62)$$

Having inverted a quantity which is of the form of a sum of the first two terms in a perturbation series in e^2, the resultant quantity is now capable of expansion to many orders in e^2. We only want the expression to order e^2 and this allows us to simplify the denominator in 3.62; the result is

$$\frac{iZ_2}{(\not{p} - m - \delta m)(1 - C(p^2))}. \quad (3.63)$$

Thus the properties of the residue and location of the pole will be as described above provided $C(p^2)$ is zero when $p^2 = m^2$. This is the property which fixes $C(p^2)$ from the finite part of $\Sigma(p)$. Proceeding as this suggests we find δm readily, and it is given by subtracting the divergences in 3.57 so that

$$\delta m = \frac{3e^2}{8\pi^2} \frac{m}{(D-4)}. \quad (3.64)$$

Things become slightly badly behaved when we attempt to calculate Z_2. Z_2 is calculated most easily by using 3.60 and the definition of $C(p^2)$. If we work at first with $p^2 \neq m^2$, so that $C(p^2)$ is present, we find in the end that

$$Z_2^{-1} = \lim_{p^2 \to m^2} \left\{ 1 + \frac{e^2}{8\pi^2(D-4)} - \frac{1}{2} \ln\left(1 - \frac{p^2}{m^2}\right) \right\}. \quad (3.65)$$

This is an unforeseen mishap; not only has Z_2 an infinity in the form of a pole at $D=4$, but there is an additional logarithmic infinity at $p^2 = m^2$, the point where Z_2 is defined. This is a breakdown of the renormalization prescription using δm and Z_2 as defined above. Fortunately this divergence, which is present even when $D \neq 4$, is not as difficult to resolve as the original infinity in Z_2. This infinity may be seen to be infrared and is due to the zero mass of the photon. This is because the threshold for the production of one electron and a number of photons coincides with the position of the mass shell pole at $p^2 = m^2$. To make this point easier to understand we compare the analytic structure, as a function of p^2, of the propagator in $\lambda \phi^4/4!$ theory with that of the electron propagator. In the former case there is only one kind of particle with mass m^2, the threshold in the propagator for the production of new particles occurs at $p^2 = 9m^2$ and is illustrated diagrammatically by Fig. 6 in Chapter 2. The properties of the propagator as a function of the complex variable p^2 are that there is a simple pole at $p^2 = m^2$ and a branch cut along the real axis beginning at $p^2 = 9m^2$. Because of the branch cut there is a discontinuity of the propagator (and hence an imaginary part) when $p^2 \geq 9m^2$. We illustrate this structure in Fig. 9.

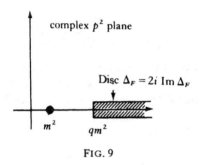

FIG. 9

For the electron propagator there is a pole in the complex p^2 plane at $p^2 = m^2$ but the corresponding cut occurs at $p^2 = (m^2 + m^2_{\text{photon}}) = m^2$. Therefore the pole sits on the lip of the cut because the mass of the photon is zero. This is the source of our difficulty since the propagator therefore has an imaginary part stretching from positive infinity in p^2 down to $p^2 = m^2$. The logarithm $\ln(1 - p^2/m^2)$ in 3.65 is thus to be expected since its presence gives the propagator an imaginary part, $\text{Im} \ln(1 - p^2/m^2) = \pi\theta(1 - p^2/m^2)$, where θ is the same step function used in Chapter 2. If we could separate the beginning of the cut from the location of the pole, then the logarithm would move with the cut and would disappear from 3.65 which would then be finite except for the simple pole at $D = 4$. We illustrate the singularity structure of the electron propagator in Fig. 10. Clearly if the photon had a small mass λ then this would solve the problem but this parameter λ would have to disappear from calculation of physical quantities. This is sometimes done, but we prefer not to do this, and wish to avoid the infrared singularity by changing the definition of Z_2. This may sound peculiar but in fact is perfectly permissible and does avoid the infrared problem. The way around this problem is to choose the momentum p^2 at which Z_2 is defined to be to the left of the electron photon threshold which starts at $p^2 = m^2$ as shown in Fig. 10. The most convenient value to the left of $p^2 = m^2$ in the complex p^2 plane is simply $p_\mu = 0$; this method of renormalization is sometimes referred to as intermediate renormalization. This new renormalization may be described as

DIMENSIONAL REGULARIZATION OF QUANTUM ELECTRODYNAMICS   113

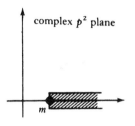

FIG. 10

follows: the inverse propagator is taken from 3.59 to be

$$-i(\not{p} - m - \Sigma). \tag{3.66}$$

The physical mass m_{phys}, and thus δm, is fixed by the condition

$$\lim_{\not{p} \to m_{\text{phys}}} -i(\not{p} - m - \Sigma) = 0 \tag{3.67}$$

but Z_2 is defined at $p_\mu = 0$ by the relation

$$\lim_{p_\mu \to 0} -i \frac{\partial}{\partial p_\mu}(\not{p} - m - \Sigma) = -iZ_2^{-1}\gamma_\mu. \tag{3.68}$$

It is a simple matter to verify that Z_2^{-1} may be defined this way and that its sole singularity is a simple pole at $D = 4$, and, of course, that δm is still given by $m_{\text{phys}} = m + \delta m$ with $\delta m = 3e^2/8\pi^2(D-4)$. With the definition of 3.68 the expression obtained for Z_2 is

$$Z_2 = 1 - \frac{e^2}{8\pi^2(D-4)}. \tag{3.69}$$

Having finally arrived at a correct way of separating the finite from the infinite parts it is now easy, by the now well established method, to incorporate counter terms into the Lagrangian L so as to eliminate these infinite parts. The modified Lagrangian together with the counter terms introduced for the vacuum polarization is

$$L + \frac{e^2 F_{\mu\nu}F^{\mu\nu}}{24\pi^2(D-4)} + \frac{ie^2\bar{\psi}\not{\partial}\psi}{8\pi^2(D-4)} + \frac{e^2 m\bar{\psi}\psi}{2\pi^2(D-4)}. \tag{3.70}$$

The third and last divergence to be eliminated is that from the vertex. We show the lowest order vertex divergence in Fig. 11.

FIG. 11

To unmask the divergence we examine the resulting Feynman integral

$$-\frac{e^3}{(2\pi)^4}\bar{u}(p+q)\int dk\, \gamma_\alpha \frac{1}{(\slashed{k}-\slashed{q}-m)}\gamma_\mu\frac{1}{(\slashed{k}-m)}\gamma^\alpha\frac{1}{(k+p)^2}u(p)$$
$$=-ie\bar{u}(p+q)\Gamma^{(2)}_\mu(p,q,p+q)u(p) \quad (3.71)$$

where $\Gamma^{(2)}_\mu(p,q,p+q)$ denotes the second order contribution to the vertex $\Gamma_\mu(p,q,p+q)$ defined in Fig. 4(b). To deal with this integral we use Feynman parameters to combine the denominators and thereby obtain

$$\bar{u}(p+q)\Gamma^{(2)}_\mu(p,q,p+q)u(p)$$
$$=\frac{ie^2}{(2\pi)^4}\Gamma(3)\int dk \int d\alpha$$
$$\cdot \int d\beta \frac{\bar{u}(p+q)\gamma_\alpha(\slashed{k}-\slashed{q}+m)\gamma_\mu(\slashed{k}+m)\gamma^\alpha u(p)}{[\alpha(k^2-m^2)+\beta\{(k-q)^2-m^2\}+(1-\alpha-\beta)(k+p)^2]^3}. \quad (3.72)$$

We continue this integral to D dimensions to render it finite and then shift the origin of the k integration in the usual way so that

$$\bar{u}(p+q)\Gamma^{(2)}_\mu(p,q,p+q)u(p)=\frac{2ie^2}{(2\pi)^4}$$
$$\int dk\, d\alpha\, d\beta \frac{\bar{u}(p+q)\gamma_\alpha(\slashed{k}-\slashed{b}-\slashed{q}+m)\gamma_\mu(\slashed{k}+\slashed{b}+m)\gamma^\alpha u(p)}{[k^2+a^2]^3}$$

where

$$a^2 = -(\alpha+\beta)m^2 - \beta(p+q)^2 - 2\beta(\alpha-1)p\cdot(p+q)$$
$$+ \alpha(1-\alpha)p^2 + \beta(q^2-p^2) \quad (3.73)$$

and

$$b = (\alpha+\beta-1)p + \beta q.$$

The numerator of 3.73 can be decomposed with the use of the various identities proved earlier in the chapter and with the further identity

$$\bar{u}(p_1)[2m\gamma_\mu - i(p_1 - p_2)^\nu \sigma_{\mu\nu}]u(p_2) = (p_1 + p_2)_\mu \bar{u}(p_1)u(p_2) \tag{3.74}$$

which follows from the use of equations of motion for the free particle spinors $\bar{u}(p_1)$, $u(p_2)$. When this is done, two independent scalar form factors, $F(q^2)$, $G(q^2)$, specify the vertex. These form factors are defined by the equation

$$\bar{u}(p+q)\Gamma_\mu(p, q, p+q)u(p) = \bar{u}(p+q)[F\gamma_\mu + q^\nu \sigma_{\mu\nu} G]u(p). \tag{3.75}$$

We shall confine our attention for the moment to the calculation of $F(q^2)$ rather than $G(q^2)$ which is in fact finite. To calculate the infinite pole part of $F(q^2)$ is quite straightforward; the result is simply

$$-\frac{e^2}{8\pi^2(D-4)}. \tag{3.76}$$

Having proceeded this far we may now check the Ward identity. To do this we take the infinite part of $-iS_F^{-1}(p)$ from 3.57, namely

$$\frac{-ie^2 \not{p}}{8\pi^2(D-4)}. \tag{3.77}$$

We find immediately that the Ward identity $\partial S_F^{-1}/\partial p_\mu = \Gamma_\mu(p, 0, p)$ is satisfied for the infinite parts of 3.76, 3.77 verifying that these quantities are indeed related. However, the Ward identity brings some difficulties with it in the shape of the infrared divergence of the inverse propagator. Because of the Ward identity and the infrared divergence of $S_F^{-1}(p)$ at $p^2 = m^2$, $\Gamma_\mu(p, q, p+q)$ is infrared divergent at $p^2 = (p+q)^2 = m^2$. The content of the infrared divergence in $S_F^{-1}(p)$ was to introduce an infinity into the definition of Z_2 associated with $p^2 \to m^2$; the only way around this was to change the definition of Z_2. A similar thing occurs here; Z_1, the vertex renormalization constant defined in 3.17, has an infinity when $D = 4$ and a logarithmic infinity when $p^2 \to m^2$. This failure of Z_1 to be finite when $D \neq 4$ has, as we would expect from the analysis of $S_F^{-1}(p)$, its origin in the failure of the 'finite part' of Γ_μ to be finite when $p^2 \to m^2$. Our use of the equations of motion to derive the identity 3.74 assumes $p^2 = m^2$. If however we do not use 3.74, we may work out the finite part, $\Gamma_\mu(p, q, p+q)$, for arbitrary q and $p^2 \neq m^2$. The infrared divergence that we are dealing with occurs only in $F(q^2)$ so that there is no need to work out $G(q^2)$ for $p^2 \neq m^2$. The actual evaluation of the integral over the Feynman parameters for the finite part is lengthy and tedious but it can be done; the resultant expression involves one transcendental function, the Spence function. The infrared divergence may be tackled in two ways; one way is the one referred to above where one calculates Γ_μ for off-mass shell electrons; a second way is to give the photon a mass Λ by replacing the photon propagator $-ig_{\mu\nu}/p^2$ by $-ig_{\mu\nu}/(p^2-\Lambda^2)$. The infrared divergence then appears as $\Lambda \to 0$. We used the former method when calculating the electron propagator; we shall use the latter method for the vertex. The finite part of $F(q^2)$ is found to be†

$$F(q^2) = \frac{e^2}{8\pi^2} \left[\text{const.} + \frac{1+a^2}{1-a^2} \left\{ 2\ln a \ln(1+a) - \frac{\ln^2 a}{2} - 2f(a) + \frac{\pi^2}{6} \right\} \right.$$
$$\left. + \frac{(3a^2+2a+3)}{2(a^2-1)} \ln a + \frac{(1+a^2)}{(1-a^2)} \ln a \ln\left(\frac{m^2}{\Lambda^2}\right) \right] \quad (3.78)$$

where

$$a = 1 - \frac{q^2}{2m^2} + \frac{q^2}{m^2}\left(\frac{1}{4} - \frac{m_2}{q^2}\right)^{1/2}$$

The infrared divergence is seen to occur in the last term and is logarithmic in nature.‡ $G(q^2)$ is infrared finite and is given by

$$G(q^2) = \frac{e^2}{4\pi^2} \frac{a \ln a}{a^2-1}. \quad (3.79)$$

To define Z_1, the vertex renormalization constant, the intermediate renormalization scheme used to define Z_2 suggests that instead of defining it in terms of $\Gamma_\mu(p, 0, p)$ at $p^2 = m^2$ we define it in terms of $\Gamma_\mu(p, 0, p)$, $p = 0$. With all the momenta set equal to zero, Γ_μ is simply $C\gamma_\mu$ where C is an (infinite) constant. This infinite constant C has in the usual way an undetermined finite part. This finite part is fixed by the choice of subtraction constant which we shall choose to be

† In 3.78 $f(a)$ is the Spence function and is defined by

$$f(a) = \int_0^a dx \frac{\ln(1+x)}{x}.$$

‡ The physical cross-sections will be shown in Chapter 4 to be independent of Λ thus justifying the use of this mathematical device.

$+e^2/(8\pi^2(D-4))$; this makes C have no finite part (compare with 3.76). The definition which we then give Z_1 is

$$\Gamma_\mu(p, q, p+q) = Z_1^{-1}\gamma_\mu, \qquad p = q = 0. \tag{3.80}$$

This means that we can straight away deduce that

$$Z_1 = 1 - \frac{e^2}{8\pi^2(D-4)} = Z_2. \tag{3.81}$$

We have verified the Ward identity as we wished, though in doing so we had to change the definitions of Z_1 and Z_2 because of the infrared problem. We shall take up this problem again in the next chapter.

The final form of the modified Lagrangian with all counter terms present is

$$L + \frac{e^2}{24\pi^2(D-4)}F_{\mu\nu}F^{\mu\nu} + \frac{ie^2}{8\pi^2(D-4)}\bar{\psi}\slashed{\partial}\psi + \frac{e^2 m\bar{\psi}\psi}{2\pi^2(D-4)} - \frac{e^3\bar{\psi}\gamma^\mu\psi A_\mu}{8\pi^2(D-4)}. \tag{3.82}$$

For convenience we give the relations between the subtraction constants, the renormalization constants and the mass counter term.

$$Z_1 = Z_2 = 1 - \frac{e^2}{8\pi^2(D-4)}$$

$$Z_3 = 1 + \frac{e^2}{6\pi^2(D-4)} \tag{3.83}$$

$$m_R = m + \delta m, \qquad \delta m = \frac{3e^2 m}{8\pi^2(D-4)}.$$

For the final part of this section we remind the reader that, as in $\lambda\phi^4/4!$ theory the counter terms given in 3.82 must be multiplied by the dummy parameter S, and renormalization performed order by order in S and S set equal to unity at the end. Further, we introduce the renormalized fields so that we may write out the renormalized Lagrangian. As in the last chapter the Lagrangian L is the bare Lagrangian L_0 and is related to L_R, the renormalized Lagrangian by

$$\begin{aligned}L_0 = L_R &+ (Z_3 - 1)(-\tfrac{1}{2}(\partial_\lambda A_R^\lambda)^2) + (Z_3 - 1)(-\tfrac{1}{4}F_{\mu\nu}F^{\mu\nu}) \\ &+ (Z_2 - 1)(\bar{\psi}_R i\slashed{\partial}\psi_R - m_R\bar{\psi}_R) + Z_2\,\delta m\,\bar{\psi}_R\psi_R \\ &- (Z_1 - 1)e_R\bar{\psi}_R\gamma^\mu\psi_R A_{\mu R}.\end{aligned} \tag{3.84}$$

where the usual relations exist between the bare and the renormalized quantities, namely

$$A_0^\mu(x) = Z_3^{-1/2} A_R^\mu$$
$$\psi_0(x) = Z_2^{1/2} \psi_R(x),$$
$$e_0 = \frac{Z_1}{Z_2} Z_3^{1/2} e_R = Z_3^{-1/2} e_R, \qquad (3.85)$$
$$m_R = m + \delta m.$$

The Lagrangian L_R in which all quantities are the renormalized ones gives finite results for all matrix elements; all infinities are cancelled by the renormalization constants whose magnitudes have no direct physical consequences.

Shortcoming of the Dimensional Method: The ε Symbol and the Triangle Anomaly

So far we have not referred to any shortcoming of the dimensional regularization method. It would be unwise to allow a false impression to be created in this regard. Therefore we wish to describe the main shortcoming of the dimensional method. Fortunately this shortcoming is limited in its occurrence to a few special situations, one of which we shall describe. The physical process involved is the electromagnetic decay of the neutral π meson, π^0. The decay is into two (real) photons

$$\pi^0 \to \gamma + \gamma \qquad (3.86)$$

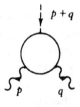

FIG. 12

The decay is represented in Fig. 12, the dotted line is the π^0 and the wiggly lines are the photons. The matrix element for the decay is

$$\varepsilon_{\mu\nu\alpha\beta} \varepsilon_p^\mu \varepsilon_q^\nu p^\alpha q^\beta M(p \cdot q), \qquad (3.87)$$

M is an invariant scalar function of the only momentum variable, $p \cdot q$; ε_p^μ and ε_q^ν are the polarization vectors of the photons; the matrix element is proportional to the permutation symbol $\varepsilon_{\mu\nu\alpha\beta}$ because of the negative parity of the π^0. A popular method for calculating $M(p \cdot q)$ is to assume (with some justification, but we will not go into it here) that the matrix element of the π^0 field operator $\pi^0(x)$ is proportional to the matrix element of the divergence of an axial vector current

$$\langle A|\pi^0(x)|B\rangle = C\langle A|\partial_\mu j^{\mu 5}(x)|B\rangle$$
$$j_\mu^5(x) = \bar\psi(x)\gamma_\mu\gamma_5\psi(x) \tag{3.88}$$

where $\psi(x)$, $\bar\psi(x)$ are some fundamental Fermions which also couple to the hadronic part of electromagnetic current

$$j_\mu(x) = \bar\psi(x)\gamma_\mu\psi(x). \tag{3.89}$$

The lowest order graph contributing to this decay is the so-called triangle graph shown in Fig. 13. The upper vertex has been labelled A

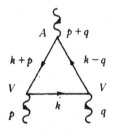

FIG. 13

because it is the axial vertex and the two lower vector vertices are correspondingly labelled V. The closed loop is made up of the Fermions $\psi(x)$ which are taken to be massless. It is also customary to calculate this graph in the limit of zero mass for the π^0, i.e. $(p+q)^2 = 2p \cdot q = 0$. (The full decay matrix element should also include the graph with the photon legs interchanged.) The Feynman integral to be evaluated is

$$(p+q)^\kappa \varepsilon_p^\mu \varepsilon_q^\nu \int \frac{dk}{(2\pi)^4} \frac{\text{Tr}\{\gamma_\mu \slashed{k}\gamma_\nu(\slashed{k}-\slashed{q})\gamma_5\gamma_\kappa(\slashed{k}+\slashed{p})\}}{k^2(k-q)^2(k+p)^2}. \tag{3.90}$$

Evaluating the numerator of 3.90 reduces the expression to the form

$$i\varepsilon_p^\mu \varepsilon_q^\nu \int \frac{dk}{(2\pi)^4}$$

$$\cdot \frac{\varepsilon_{\mu\nu\alpha\beta}[-k^\alpha q^\beta(p+q)\cdot(k+p)+k^\alpha p^\beta(k-q)\cdot(p+q)-k^\alpha(p+q)^\beta(k-q)\cdot(k+p)]}{k^2(k-q)^2(k+p)^2} \quad (3.91)$$

This integral may be seen by inspection to be linearly divergent; however, if we react as usual and try to render it finite by continuation in D we are faced with the problem of what to do about the ε symbol $\varepsilon_{\mu\nu\alpha\beta}$. The problem is that $\varepsilon_{\mu\nu\alpha\beta}$ has four indices in four dimensions, three in three etc. and has no natural unique continuation to an arbitrary dimension D. Therefore we are faced with an impasse since we cannot construct our desired continuation. An important point to realize is that in the limit of zero mass for the π^0 the axial vector current of 3.88 is conserved; there are then Ward identities for the conservation of this axial current which imply that $M(p \cdot q)$ is zero for $p \cdot q = 0$. This we find to be false by evaluation of the triangle diagram, thus the perturbative solution to the field theory violates the axial vector Ward identity. This is known as the Adler, Bell, Jackiw triangle anomaly.[4] This point provides a little more understanding of the failure of the dimensional method in this case. The reasoning goes as follows: the dimensional method preserves both axial vector and vector Ward identities in D dimensions. However, as we have seen above, the symbol $\varepsilon_{\mu\nu\alpha\beta}$ can be involved in a Ward identity and it is strictly a four-dimensional object. Thus if we attempt to continue the integral which is the coefficient of $\varepsilon_{\mu\nu\alpha\beta}$ we shall in general violate the Ward identity except when $D=4$, i.e. the breaking of the Ward identity can be written symbolically as $(D-4)I$ for some I. This would be a rather esoteric point if it were not for the fact that I is itself infinite in the case of the triangle graph, i.e. there is a simple pole at $D=4$, and thus the limit of $(D-4)I$ when $D=4$ is some finite quantity which is not zero and which violates the Ward identity. The case of the triangle anomaly is rather special; there are a few more cases of violations of the axial vector Ward identities by the same mechanism but luckily the total number of such violations is small, and finite, and can in some circumstances be totally eliminated from the theory. It is nevertheless true that in some theories of both vector and axial vector currents if the triangle anomaly is present the proof of the renormalizability by dimensional techniques is not possible.

Criterion for Renormalizability in General

Having devoted considerable space to studying the renormalization of $\lambda\phi^4/4!$ theory and QED we wish to look further afield at the renormalization of other field theories. An obvious question to which we would like an answer is, is there a way in which we can tell in advance whether a field theory defined by a given Lagrangian L is renormalizable in perturbation theory or not? It is difficult to give a watertight and complete answer to this question. But it is possible to derive a necessary condition for renormalizability which in a number of cases is also a sufficient condition. We motivate this criterion by looking at a theory which is not renormalizable and tracing the source of the non-renormalizability. The Lagrangian is given by

$$L = \tfrac{1}{2}(\partial_\mu \phi)^2 - \frac{m^2 \phi^2}{2} - \frac{\lambda}{5!}\phi^5 \qquad (3.92)$$

with Feynman rules

$$
\begin{array}{ll}
\times & -i\lambda \\
\xrightarrow{\ \ p\ \ } & \dfrac{+i}{p^2 - m^2 + i\varepsilon} \\
\text{loop momentum} & \displaystyle\int \dfrac{dk}{(2\pi)^4} \\
\text{symmetry factor} & S
\end{array} \qquad (3.93)
$$

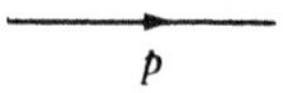

Consider the graph shown in Fig. 14 contributing to the propagator. By simple power counting we find this graph to be quartically divergent instead of the more familiar quadratic divergence of $\lambda\phi^4/4!$ theory. Now look at a higher order contribution to the propagator.

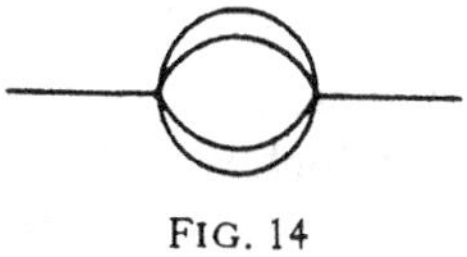

FIG. 14

Fig. 15

Power counting reveals the divergence of this graph to be much worse with a growth proportional to the sixth power of the ultraviolet momentum cut-off. It is an easy task to verify that in general the higher the order the worse the nature of the divergence. This is entirely foreign to our experience up to this point where the nature of the divergence was the same in higher orders. The trouble may be traced to the interaction term $-\lambda\phi^5/5!$. The action $S = \int dx\, L$ is a dimensionless quantity; it is therefore a simple matter to deduce that the coupling constant λ has the dimension of an inverse mass. However, the dimension of a Green's function such as the propagator is fixed, in this case to be that of inverse mass squared. Thus the higher the order in perturbation theory, i.e. the higher the powers of inverse mass multiplying the basic Feynman integral, the higher the dimension in mass units of the basic Feynman integral and the worse its divergence. This means that in each order the nature of the subtraction terms needed to eliminate the infinity changes and that in the end there are an infinite number of subtraction terms needed to make the propagator and thus any other Green's function finite. To carry this procedure through we would need an infinite number of renormalized quantities in which to absorb this infinity of subtraction terms or renormalization constants. But since the renormalized quantities all have a physical significance this clearly leads to a rather worthless physical theory. Such theories are called non-renormalizable. Having traced the source of the trouble, namely the dimension of λ, we seek a more general method of stating it. This is provided by the following; write

$$L = \tfrac{1}{2}(\partial_\mu \phi)^2 - \frac{m^2 \phi^2}{2} - \lambda L_I, \qquad L_I = \frac{\phi^5}{5!}, \qquad (3.94)$$

then the dimension of L_I in mass units is five and the trouble is seen to arise because dim $L_I > 4$. For if dim $L_I \leq 4$ then the coupling constant λ would either be dimensionless (dim $L_I = 4$) or have the dimension of a positive power of a mass which would not be a problem. Hence we term theories, for which the interaction term L_I in the Lagrangian satisfies dim $L_I > 4$, non-renormalizable. We now see at once that this means that $\lambda\phi^6/6!$, $\lambda\phi^7/7!$ etc. are all non-renormalizable theories. An example of a non-renormalizable theory involving Fermions would be one where $L_I = (\bar\psi\psi)^2$ which has dimension six. The renormalizable theories that we considered previous to this section, $\lambda\phi^4/4!$ and QED, both have dim $L_I = 4$, i.e. a dimensionless coupling constant. The last thing to do is to examine a theory for which dim $L_I < 4$. An example of this is provided by the Lagrangian

$$L = \tfrac{1}{2}(\partial_\mu\phi)^2 - \frac{m^2\phi^2}{2} - \frac{\lambda\phi^3}{3!} \qquad (3.95)$$

with Feynman rules

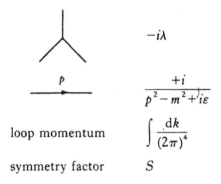

If we examine the lowest order propagator factor graph of Fig. 16, we find by power counting that it is only logarithmically divergent instead

FIG. 16

of quadratically divergent. A higher order contribution such as that shown in Fig. 17 is finite by power counting. This is what we expect,

FIG. 17

the dimension of the coupling constant now being positive rather than negative tends to improve the divergence situation. Another divergence is found to be absent when we examine the three-particle vertex

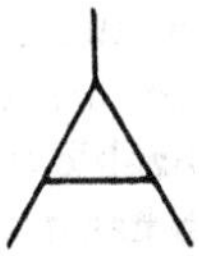

FIG. 18

of Fig. 18, which we would expect to be logarithmically divergent. Power counting reveals this graph to be convergent. In fact there is only one source of divergence in $\lambda\phi^3/3!$ theory and that is the divergence of the propagator. Theories such as $\lambda\phi^3/3!$ theory where the number of independent divergences is reduced are called super-renormalizable. In summary we can say that if dim $L_I < 4$ a theory is super-renormalizable, if dim $L_I > 4$ it is non-renormalizable, and if dim $L_I = 4$ we expect, but cannot be sure, that it is renormalizable.

References

1. J. D. Bjorken and S. D. Drell, 'Relativistic Quantum Mechanics', McGraw-Hill (1964).
2. G. 't Hooft and M. Veltman, *Nucl. Phys.*, B**50**, 318 (1972).
3. G. 't Hooft and M. Veltman, *Diagrammar CERN* (1973).
4. S. Adler, *in* 'Lectures on Elementary Particles and Quantum Field Theory' (S. Deser, M. Grisaru and H. Pendleton, eds), M.I.T. Press (1970).

FOUR

The Gauge and Infrared Properties of Quantum Electrodynamics

Gauge Fixing Terms

In this chapter we want to do two things: the first is to discuss the gauge properties of QED more completely; the second is to give an account of the infrared problems of QED and their resolution. The important technical device which is so relevant for discussing the gauge properties of QED is the *gauge fixing term*. This name is usually applied to the expression $-\frac{1}{2}(\partial_\lambda A^\lambda)^2$ which appears in the QED Lagrangian of eq. 3.1 in Chapter 3. In classical physics, treatments of Maxwell's equations are usually conducted without ever mentioning such a term. Let us now suppose that the gauge-fixing term is absent and try to understand why it is needed. Consider the QED Lagrangian with no interaction, i.e. with the electron term absent so that we have

$$L = -\tfrac{1}{4}(\partial_\mu A_\nu - \partial_\nu A_\mu)^2 = -\tfrac{1}{4}F_{\mu\nu}F^{\mu\nu}. \tag{4.1}$$

This Lagrangian can easily be seen to possess the well-known gauge invariance property: namely that L is unchanged if A_μ is replaced by $A_\mu + \partial_\mu \Lambda$ for some function Λ. Now the physical content of the theory is that the solution of the Euler–Lagrange equations for L is the radiation field. But the gauge invariance property means that A_μ and $A_\mu + \partial_\mu \Lambda$ are equally good as radiation fields. Mathematically speaking this means that given a set of initial data for the radiation field there is no unique solution to the Euler–Lagrange problem. This property is, of course, noted in classical (i.e. non-quantum) treatments of Maxwell's equations and a gauge condition is usually introduced. For example, in the Coulomb or radiation gauge we require

$$\nabla \cdot \mathbf{A} = 0, \qquad A_0 = 0. \tag{4.2}$$

This means that if we start with an arbitrary solution A'_μ to the Euler–Lagrange equations, and then change A'_μ to $A'_\mu + \partial_\mu \Lambda = A_\mu$ so that,

$$\nabla \cdot \mathbf{A} = \nabla \cdot \mathbf{A}' + \nabla^2 \Lambda = 0, \tag{4.3}$$

then Λ is given by[†]

$$\Lambda = \frac{-1}{4\pi} \int d^3x' \frac{\nabla \cdot \mathbf{A}'}{|\mathbf{x} - \mathbf{x}'|}. \tag{4.4}$$

Stated in a more physical way, the non-uniqueness of gauge invariance is telling us that not all the degrees of freedom of the radiation field are independent. In fact, of the four components of A_μ only two are independent. We notice that the radiation gauge imposes two linear relations among the field components A_μ, leaving two. This is of paramount importance when quantizing the radiation field, as when quantizing a classical system one must only quantize independent degrees of freedom. As far as the physical content of the theory is concerned there will be many different, but physically equivalent, gauge conditions possible. Each has its own advantages; the choice of gauge condition is made only when the particular technical features of a calculation which can be simplified by its introduction are known. The radiation gauge has the disadvantage that it is not manifestly covariant. It has the advantage that the two independent degrees of freedom of A_μ are both transverse. In quantizing the radiation field the problem of gauge invariance is tackled by going back to the original Lagrangian and adding in the gauge fixing term $-\frac{1}{2}(\partial_\lambda A^\lambda)^2$. The presence of this term means that the Lagrangian is no longer invariant under the group of transformations $A_\mu \to A_\mu + \partial_\mu \Lambda$. The resulting Euler–Lagrange equations are different and one can obtain a unique solution to these equations. It is not immediately obvious in the quantum approach that the physics is unaltered by this

---

[†] A_0 is made zero by first choosing a $\tilde{\Lambda}$ such that $\tilde{\Lambda} = -\int dx_0\, A'_0$ and then choosing the Λ of 4.4. This is consistent if $\partial \Lambda / \partial x_0 = 0$.

modification of the Lagrangian. However it can be seen quite directly using the functional integral definition of the quantized theory given in Chapter 1. We shall work with the functional $Z[J]$, which generates the S-matrix elements in coordinate space. Remember that we have not yet introduced Fermi fields so that Z takes the simple form

$$Z[J_\mu] = \frac{\int \mathcal{D}[A_\mu] \exp\left[i \int dx \left\{-\tfrac{1}{4}F_{\mu\nu}F^{\mu\nu} - \tfrac{1}{2}(\partial_\lambda A^\lambda)^2 + J_\mu A^\mu\right\}\right]}{\int \mathcal{D}[A_\mu] \exp\left[i \int dx \left\{-\tfrac{1}{4}F_{\mu\nu}F^{\mu\nu} - \tfrac{1}{2}(\partial_\lambda A^\lambda)^2\right\}\right]}.$$
(4.5)

We now take a step back, by deleting the gauge fixing term and define the functional $\bar{Z}[J]$ by

$$\bar{Z}[J_\mu] = \frac{\int \mathcal{D}[A_\mu] \exp\left[i \int dx \left\{-\tfrac{1}{4}F_{\mu\nu}F^{\mu\nu} + J_\mu A^\mu\right\}\right]}{\int \mathcal{D}[A_\mu] \exp\left[i \int dx \left\{-\tfrac{1}{4}F_{\mu\nu}F^{\mu\nu}\right\}\right]}.$$
(4.6)

We now modify the functional $\bar{Z}[J]$ in a way which we first describe and then explain why it gets round the gauge problem. We add to the action the term $F[A, \Lambda]$ where $F[A, \Lambda]$ satisfies

$$\int \mathcal{D}[\Lambda] \exp[iF[A_\Lambda]] = 1$$
(4.7)

and F is *not* invariant under the group of gauge transformations $A_\mu \to A_\mu + \partial_\mu \Lambda$. The symbol $\mathcal{D}[\Lambda]$ in 4.7 stands for integration over the gauge functions Λ. Let us now be specific. We choose

$$F = \int dx \left\{-\tfrac{1}{2}(\partial_\lambda A^\lambda)^2\right\} + G,$$
(4.8)

where G is defined by

$$G = i \ln \int \mathcal{D}[\Lambda] \exp\left[i \int dx \left\{-\tfrac{1}{2}(\partial_\lambda A_\Lambda^\lambda)^2\right\}\right]$$
(4.9)

i.e.

$$\exp[-iG] = \int \mathcal{D}[\Lambda] \exp\left[i \int dx \left\{-\tfrac{1}{2}(\partial_\lambda A_\Lambda^\lambda)^2\right\}\right]$$
(4.10)

and A_Λ^μ is the gauge-transformed field $A^\mu + \partial^\mu \Lambda$. This choice of F satisfies our requirements. F is not a gauge-invariant quantity because of the presence of the term $-\tfrac{1}{2}(\partial_\lambda A^\lambda)^2$. Also condition 4.7 is automatically satisfied when G is chosen as in 4.9,10. We call the functional which results when the term F is added to the action, $\tilde{Z}[J_\mu]$. Hence

$$\tilde{Z}[J_\mu] = \frac{\int \mathcal{D}[A_\mu] \exp\left[i \int dx \{-\tfrac{1}{4}F_{\mu\nu}F^{\mu\nu} + J_\mu A^\mu\} + iF\right]}{\int \mathcal{D}[A_\mu] \exp\left[i \int dx \{-\tfrac{1}{4}F_{\mu\nu}F^{\mu\nu}\} + iF\right]} \quad (4.11)$$

$$= \frac{\int \mathcal{D}[A_\mu] \exp\left[i \int dx \{-\tfrac{1}{4}F_{\mu\nu}F^{\mu\nu} - \tfrac{1}{2}(\partial_\lambda A^\lambda)^2 + J_\mu A^\mu\} + iG\right]}{\int \mathcal{D}[A_\mu] \exp\left[i \int dx \{-\tfrac{1}{4}F_{\mu\nu}F^{\mu\nu} - \tfrac{1}{2}(\partial_\lambda A^\lambda)^2\} + iG\right]}. \quad (4.12)$$

The term G may be calculated from 4.10. Note that

$$-\tfrac{1}{2}(\partial_\lambda A^\lambda_\Lambda)^2 = -\tfrac{1}{2}(\partial_\lambda A^\lambda + \Box \Lambda)^2. \quad (4.13)$$

The change of variable $\Lambda \to \bar{\Lambda}$ is suggested, with $\bar{\Lambda}$ given by

$$\Lambda(x) = -\int dx' \, D_F(x - x')\{-\partial_\lambda A^\lambda(x') + \bar{\Lambda}(x')\} \quad (4.14)$$

where $D_F(x - y)$ is the Green's function introduced in Chapter 1 and satisfies

$$\Box D_F(x - y) = -\delta(x - y). \quad (4.15)$$

We can now write, using 4.10,

$$\exp[-iG] = D \int \mathcal{D}[\bar{\Lambda}] \exp\left[i \int dx \left\{-\frac{\bar{\Lambda}^2}{2}\right\}\right] \quad (4.16)$$

where D is the Fredholm determinant of the integral transformation (4.14). We now normalize the measure Λ so that

$$\int \mathcal{D}[\Lambda] \exp\left[i \int dx \left\{-\frac{\bar{\Lambda}^2}{2}\right\}\right] = 1, \quad (4.17)$$

therefore

$$\exp[-iG] = D, \quad (4.18)$$

but we know from Chapter 1 that the Fredholm determinant D is given by the formula

$$D = \exp\left[\int dx \frac{K(x, y)}{2}\right] \quad (4.19)$$

where $K(x, y)$ is the kernel of the integral equation. In this case

$$K(x, y) = D_F(x - y) \quad (4.20)$$

GAUGE AND INFRARED PROPERTIES 129

so
$$D = \exp\left[\int dx \frac{D_F(0)}{2}\right]. \qquad (4.21)$$

Now we have trouble for $D_F(0) = \infty$. Nevertheless we ask the forebearance of the reader and proceed as though this did not matter. We shall anyway be arguing that this term is not physically relevant. Equation 4.18 implies that

$$\exp[iG] = D^{-1} = \exp\left[-\int dx \frac{D_F(0)}{2}\right]. \qquad (4.22)$$

We would like to write D^{-1} in another way. Using the results of Chapter 1 we can show that

$$\begin{aligned} D^{-1} &= \int \mathscr{D}[\Lambda] \exp\left[i \int dx \frac{\Lambda \Box \Lambda}{2}\right] \\ &= \int \mathscr{D}[\Lambda] \exp\left[-i \int dx \frac{\{\partial_\mu \Lambda\}^2}{2}\right] \end{aligned} \qquad (4.23)$$

where the last step is accomplished by making use of the fact that a four-divergence does not contribute to the action integral.† The functional $\tilde{Z}[J]$ now takes on the form

$$\int \mathscr{D}[A_\mu]\mathscr{D}[\Lambda] \exp\left[i\int dx\left\{-\tfrac{1}{4}F_{\mu\nu}F^{\mu\nu} - \tfrac{1}{2}(\partial_\lambda A^\lambda)^2 - \frac{(\partial_\mu \Lambda)^2}{2} + J_\mu A^\mu\right\}\right]/A \qquad (4.24)$$

where A stands for the denominator. This form would be the same as that used in Chapter 1, (except, of course, that we have not yet included the Fermi fields), if it were not for the presence of the free scalar field Λ. However we note that Λ is a free field and so does not interact with the photon; also that its action appears in 4.24 with the wrong sign in front of the derivatives. Because the field is free we disregard it in calculating Feynman diagrams; it corresponds to a phase only, although as we have seen this phase is infinite. The presence of the wrong sign in the action leads to it being called a "ghost". It means that the Feynman rules for such a particle carry a minus sign for each closed loop; in this respect it is like a Fermion. Summarizing what we have done, we say that the addition of the term

† We are, of course, assuming that Λ and $\partial \Lambda/\partial x$ vanish at ∞.

F to the action amounts to the introduction of the gauge-fixing term and a harmless free ghost scalar field Λ.† We have therefore solved the gauge problem. However we must answer the question would the physics change if a different choice was made for F. When introducing F we merely required it to break the gauge invariance of the original action and to satisfy the integrability condition 4.7. Suppose that we have two different F's, F_1 and F_2 which satisfy both these requirements. We show that they both give rise to the same S-matrix. To see this consider the integral below

$$K[J] = \int \mathscr{D}[A_\mu]\mathscr{D}[\Lambda]$$
$$\exp\left[i\int dx\, \{-\tfrac{1}{4}F_{\mu\nu}F^{\mu\nu} + J_\mu A^\mu\} + iF_1[A] + iF_2[\dot{A}_\Lambda]\right]. \quad (4.25)$$

The relation we need is obtained by evaluating this integral in two ways. If we perform first the integral with respect to Λ then we obtain, using 4.7,

$$K_1[J] = \int \mathscr{D}[A_\mu] \exp\left[i\int dx\, \{-\tfrac{1}{4}F_{\mu\nu}F^{\mu\nu} + J_\mu A^\mu\} + iF_1[A]\right]. \quad (4.26)$$

Now we change variable from A_μ to $\bar{A}_\mu$ where $\bar{A}_\mu$ is defined by

$$\bar{A}^\mu = A^\mu_\Lambda = A^\mu + \partial^\mu \Lambda, \quad (4.27)$$

evidently

$$\mathscr{D}[\bar{A}_\mu] = \mathscr{D}[A_\mu]. \quad (4.28)$$

Remembering that the action is gauge invariant we find that

$$K[J] = \int \mathscr{D}[A_\mu]\mathscr{D}[\Lambda]$$
$$\cdot \exp\left[i\int dx\, \{-\tfrac{1}{4}F_{\mu\nu}F^{\mu\nu} + J_\mu A^\mu_{\Lambda^{-1}}\} + iF_1[A_{\Lambda^{-1}}] + iF_2[A]\right]$$
$$= \int \mathscr{D}[A_\mu]\mathscr{D}[\Lambda]$$
$$\cdot \exp\left[i\int dx\, \{-\tfrac{1}{4}F_{\mu\nu}F^{\mu\nu} + J_\mu A^\mu_\Lambda\}\right.$$
$$\left. + iF_1[A_\Lambda] + iF_2[A]\right] \quad (4.29)$$

†QED is usually discussed without any reference to ghost particles. In theories with a non-Abelian gauge symmetry ghost particles cannot be avoided. The method we have used to treat the gauge problem of QED intentionally parallels that used for non-Abelian gauge theories.[1]

GAUGE AND INFRARED PROPERTIES 131

where in the last step of 4.29 we changed variable from Λ^{-1} to Λ. Consider the term in the exponential of 4.29 which involves the source J_μ, it is of the form

$$\int dx\, J_\mu A^\mu_\Lambda = \int dx\, J_\mu (A^\mu + \partial^\mu \Lambda). \tag{4.30}$$

But the term $\int dx\, (J_\mu A^\mu + \partial_\mu J^\mu \Lambda)$ differs from this only by a four-divergence which contributes zero to the action integral, hence

$$\int dx\, J_\mu A^\mu_\Lambda = \int dx\, J_\mu A^\mu + \int dx\, \partial_\mu J^\mu \Lambda. \tag{4.31}$$

Now we require, for the purposes of this derivation, the source function for the photon to be conserved, or transverse, i.e. $\partial_\mu J^\mu = 0$, and in this case we have

$$\int dx\, J_\mu A^\mu_\Lambda = \int dx\, J_\mu A^\mu. \tag{4.32}$$

In other words the source term in the action is gauge-invariant for a transverse source. With this information we see that if we again use 4.7 $K[J]$ may be written

$$K[J] = \int \mathcal{D}[A_\mu] \exp\left[i \int dx\, \{-\tfrac{1}{4} F_{\mu\nu} F^{\mu\nu} + J_\mu A^\mu\} + i F_2[A]\right]$$

$$= K_2[J]. \tag{4.33}$$

The result of our two integration exercises then is simply

$$K_1[J] = K_2[J]. \tag{4.34}$$

This has the consequence that the functional integral which is the numerator of $\tilde{Z}[J]$ may be performed with $F_1[A]$ or $F_2[A]$ in place of F with the same result. A similar property holds for the denominator. $\tilde{Z}[J]$ is independent of the specific choice of $F[A]$; since $\tilde{Z}[J]$ can be written in the form given in 4.24 we know that it generates S-matrix elements as shown in Chapter 1. The S-matrix elements are therefore independent of F. Before we can leave this argument we must rectify the defect of not including the Fermion term in the action. To do this it is necessary to prove 4.34 with each $K[J]$, including the Fermion and the relevant source terms in the action. Consider the integral

$$K[\zeta, \bar{\zeta}, J] = \int \mathscr{D}[A]\mathscr{D}[\psi]\mathscr{D}[\bar{\psi}]\mathscr{D}[\Lambda] \exp\left[i\int dx \{-\tfrac{1}{4}F_{\mu\nu}F^{\mu\nu} \right.$$
$$\left. +\bar{\psi}(i\rlap{/}{\partial}-m)\psi - e\bar{\psi}\gamma_\mu A^\mu \psi + J_\mu A^\mu + \bar{\zeta}\psi + \bar{\psi}\zeta\} + iF_1[A] + iF_2[A_\Lambda]\right],$$
(4.35)

integrating with respect to Λ yields

$$K[\zeta, \bar{\zeta}, J] = K_1[\zeta, \bar{\zeta}, J]$$
$$= \int \mathscr{D}[A]\mathscr{D}[\psi]\mathscr{D}[\bar{\psi}] \exp\left[i\int dx \{-\tfrac{1}{4}F_{\mu\nu}F^{\mu\nu} + \bar{\psi}(i\rlap{/}{\partial}-m)\psi \right.$$
$$\left. -e\bar{\psi}\gamma_\mu A^\mu \psi + J_\mu A^\mu + \bar{\zeta}\psi + \bar{\psi}\zeta\} + iF_1[A]\right]. \quad (4.36)$$

Before doing the second integration which gives $K_2[\zeta, \bar{\zeta}, J]$ we have something to say about the behaviour of the Fermion field under the gauge transformation $A_\mu \to A_{\mu\Lambda}$. Under a gauge transformation the Fermion term $(\bar{\psi}(i\rlap{/}{\partial}-m)\psi - e\bar{\psi}\gamma_\mu A^\mu\psi)$ is required to be invariant. This is accomplished by requiring simultaneously the transformations

$$A_\mu \mapsto A_\mu + \partial_\mu \Lambda$$
$$\psi \mapsto \psi + ie\Lambda\psi = \psi_\Lambda \quad (4.37)$$
$$\bar{\psi} \mapsto \bar{\psi} - ie\Lambda\bar{\psi} = \bar{\psi}_\Lambda.$$

Equation 4.37 gives only the infinitesimal form of the transformation for the Fermions. The correct form of the transformation is $\psi(x) \to \exp[ie\Lambda]\psi(x)$. For the photon we may rewrite the gauge transformation in the form $A_\mu(x) \mapsto U^*(\Lambda)(-i\partial_\mu - A_\mu)U(\Lambda)$, where the function $U(\Lambda) = \exp[i\Lambda]$. $U(\Lambda)$ is then seen to be an element of the group $U(1)$. (This form is also suitable for Yang–Mills theories. In Yang–Mills theories $U(\Lambda)$ belongs to some non-Abelian group, $SU(2)$ say. Also $U^*(\Lambda)$ is replaceed by the inverse matrix $U^{-1}(\Lambda)$ and $(-i\partial_\mu - A_\mu)$ is replaced by the Yang–Mills covariant derivative $D_\mu = i\partial_\mu + T \cdot A_\mu$ where the matrix T belongs to a representation of the Lie algebra of, (in this case), $SU(2)$, and $U(\Lambda)$ takes the form $U(\Lambda) = \exp[iT \cdot \Lambda]$.)

GAUGE AND INFRARED PROPERTIES

With this preliminary it is now possible to show that

$$K[\zeta, \bar{\zeta}, J] = K_2[\zeta, \bar{\zeta}, J]$$
$$\cdot \int \mathcal{D}[A_\mu]\mathcal{D}[\psi]\mathcal{D}[\bar{\psi}]\mathcal{D}[\Lambda] \exp\left[i \int dx \{-\tfrac{1}{4}F_{\mu\nu}F^{\mu\nu} + \bar{\psi}(i\slashed{\partial} - m)\psi\right.$$
$$-e\bar{\psi}\gamma_\mu A^\mu \psi + J_\mu A^\mu + \bar{\zeta}\psi + ie\Lambda\bar{\zeta}\psi + \bar{\psi}\zeta - ie\Lambda\bar{\psi}\zeta\}$$
$$\left. + iF_2[A] + iF_1[A]\right]. \tag{4.38}$$

If it were not for the presence of the Fermion terms $ie\Lambda\bar{\zeta}\psi$ and $-ie\Lambda\bar{\psi}\zeta$ in 4.38 the same argument about the independence of the S-matrix on F could be given. We can regard 4.38 as merely redefining the sources ζ and $\bar{\zeta}$ to be ξ and $\bar{\xi}$ respectively where

$$\xi = (1 + ie\Lambda)\zeta$$
$$\bar{\xi} = (1 - ie\Lambda)\bar{\zeta}. \tag{4.39}$$

This suggests that we require the source functions ζ and $\bar{\zeta}$ to transform under gauge transformations according to 4.39. To be explicit if $A^\mu \to A^\mu_\Lambda$ then

$$\zeta \mapsto \zeta_\Lambda = \zeta + ie\Lambda\zeta$$
$$\bar{\zeta} \mapsto \bar{\zeta}_\Lambda = \bar{\zeta} - ie\Lambda\bar{\zeta}. \tag{4.40}$$

Having done this we define a new function $F[A, \psi, \zeta]$ by

$$F[A, \psi, \zeta] = F[A] + \int dx(\bar{\zeta}\psi + \bar{\psi}\zeta) \tag{4.41}$$

and replace the condition 4.7 by

$$\int \mathcal{D}[\Lambda] \exp\left[iF[A_\Lambda, \psi_\Lambda, \zeta_\Lambda]\right] = 1. \tag{4.42}$$

Now equation 4.38 will simplify to

$$K[\zeta, \bar{\zeta}, J] = K_2[\zeta, \bar{\zeta}, J]$$
$$= \int \mathcal{D}[A]\mathcal{D}[\psi]\mathcal{D}[\bar{\psi}] \exp\left[i\int dx \{-\tfrac{1}{4}F_{\mu\nu}F^{\mu\nu}\right.$$
$$+ \bar{\psi}(i\slashed{\partial} - m)\psi - e\bar{\psi}\gamma_\mu A^\mu \psi$$
$$\left. + J_\mu A^\mu + \bar{\zeta}\psi + \bar{\psi}\zeta\} + iF_2\right] = K. \tag{4.43}$$

The equality $K_1 = K_2$ now can be used to invoke the same argument as was used above for the gauge independence of the S-matrix. We have learned from this result that photons are emitted by gauge-invariant sources J_μ; but that Fermions are emitted by gauge-dependent sources ζ and $\bar\zeta$. Physically speaking this is because the electrons carry a charge but the photon does not. Now we have understood why the gauge fixing term is needed and seen that the gauge-fixing procedure is not unique. We have also noted that this procedure introduces unphysical ghost[1] particles which do not interact with anything and have no effect on the S-matrix. The particular choice that we shall make for the function $F[A, \psi, \zeta]$ is

$$F[A, \psi, \zeta] = -\frac{1}{2\xi}(\partial_\lambda A^\lambda)^2 + \bar\zeta\psi + \bar\psi\zeta + G. \tag{4.44}$$

The parameter ξ is included to allow us to exploit the freedom of the gauge-independence of the S-matrix by varying ξ. The generating functional $Z[\zeta, \bar\zeta, J_\mu]$ that is given by the expression

$$Z[\zeta, \bar\zeta, J] = \int \mathcal{D}[A]\mathcal{D}[\psi]\mathcal{D}[\bar\psi] \exp\left[i\int dx \left\{-\tfrac{1}{4}F_{\mu\nu}F^{\mu\nu} + \bar\psi(i\slashed{\partial}-m)\psi\right.\right.$$
$$\left.\left. -e\bar\psi\gamma_\mu A^\mu\psi - \frac{1}{2\xi}(\partial_\lambda A^\lambda)^2 + \bar\zeta\psi + \bar\psi\zeta + J_\mu A^\mu\right\} + iG\right]$$
$$\cdot \left[\int \mathcal{D}[A]\mathcal{D}[\psi]\mathcal{D}[\bar\psi] \exp\left\langle i\int dx \{-\tfrac{1}{4}F_{\mu\nu}F^{\mu\nu} + \bar\psi(i\slashed{\partial}-m)\psi\right.\right.$$
$$\left.\left. -e\bar\psi\gamma_\mu A^\mu\psi - \frac{1}{2\xi}(\partial_\lambda A^\lambda)^2\} + iG\right]^{-1} \tag{4.45}$$

The ghost term $\exp[iG]$ in 4.45 is given by eqs 4.22, 23 and therefore produces the same factor in both numerator and denominator of 4.45. In this way $\exp[iG]$ cancels between numerator and denominator and so we do not have to worry about it or the fact that it is infinite. In non-Abelian gauge theories the ghost term does not cancel in this way. When the variable ξ of 4.44 is unity $Z[\zeta, \bar\zeta, J_\mu]$ is identical with that used in Chapter 1. For other values of ξ the Feynman diagrams derived from Z will depend on ξ. Let us see the form that this dependence takes. There is only one place where ξ appears in the Feynman rules and this is in the photon propagator. The photon propagator for a general value of ξ is

$$\left(g_{\mu\nu} - \frac{q_\mu q_\nu}{q^2}\right)\frac{-i}{q^2} - i\xi\frac{q_\mu q_\nu}{q^4}. \tag{4.46}$$

When $\xi = 1$ we regain the propagator $-ig_{\mu\nu}/q^2$ that we have previously worked with. An important fact is now obvious. It is that when the gauge fixing term is absent, (i.e. $\xi = \infty$), the photon propagator does not exist since the longitudinal part becomes infinite. This is a direct way of seeing that a gauge fixing term is necessary for the quantized theory. When calculations of S-matrix elements are made in QED the total S-matrix elements obtained by adding together all allowed graphs of the same order in e cannot depend on ξ. Nevertheless particular calculations can often be simplified by choosing ξ at the outset to be a certain fixed value. The two most common choices for ξ are 0 and 1. These values have names $\xi = 1$ is called the Feynman gauge and $\xi = 0$ is called the Landau gauge. The main advantage of the Feynman gauge is that the propagator takes on its simplest form thus speeding up the calculation. The main advantage of the Landau gauge is that the photon propagator is purely transverse. Therefore it contributes zero when contracted with a momentum q_μ coming from elsewhere in the expression for the Feynman diagram. It should also be realized that, although the Feynman rules only contain ξ in the photon propagator; the electron propagator and the electron–photon vertex will also depend on ξ in higher orders of perturbation theory. This is simply because internal photon lines occur in the higher order contributions to these Green's functions.

The Ward Identities

In the last chapter we used the Ward identity in proving the equality $Z_1 = Z_2$ for the renormalization constants. We are now in a position to derive this relation and to show that it is just one of many relations which follow from the gauge invariance of the theory. The convenient starting point from which to derive these relations is the expression 4.45 for the functional $Z[\zeta, \bar\zeta, J]$. We take this expression and make use of the fact that it remains unchanged if we change the integration variables from A^μ, ψ, $\bar\psi$ to A^μ_Λ, ψ_Λ, $\bar\psi_\Lambda$. For our own mathematical convenience, however, we do this only in the integral appearing in the numerator of 4.45. The terms in the integrand which change give rise to the additional factor

$$\exp\left[i\int dx\left\{-\frac{1}{2\xi}(2\partial_\lambda A^\lambda \Box \Lambda + O(\Lambda^2)) + ie\Lambda(\bar\zeta\psi - \bar\psi\zeta) + J_\mu \partial^\mu \Lambda\right\}\right]$$
(4.47)

where we have expanded terms to first order in the gauge transformation Λ. Expanding the exponential to first order in Λ gives the expression

$$1 + i \int dy \left\{ -\frac{1}{\xi}\partial_\lambda A^\lambda \Box \Lambda(y) + ie\Lambda(y)(\bar{\zeta}\psi - \bar{\psi}\zeta) + J_\mu \partial^\mu \Lambda(y) \right\}. \quad (4.48)$$

But since $Z[\zeta, \bar{\zeta}, J]$ is unchanged by this change of variable the coefficient of Λ in this additional linear term in Λ must functionally integrate to zero for arbitrary Λ. In other words we have the equation

$$\int \mathcal{D}[A]\mathcal{D}[\psi]\mathcal{D}[\bar{\psi}]\left\{ -\frac{\Box}{\xi}\partial_\lambda A^\lambda(y) + ie(\bar{\zeta}(y)\psi(y) - \bar{\psi}(y)\zeta(y)) - \partial_\lambda J^\lambda(y) \right\}$$

$$\exp\left[iS[A, \psi, \bar{\psi}] + i\bar{\zeta}\psi + i\bar{\psi}\zeta + J_\mu A^\mu\right]/A = 0, \quad (4.49)$$

where $S[A, \psi, \bar{\psi}]$ stands for the action, A is the denominator term, and we have used integration by parts to remove the derivatives from $\Lambda(y)$. It is a simple matter to rewrite 4.49 as a functional differential equation for $Z[\zeta, \bar{\zeta}, J]$. The equation that we obtain on performing this rewriting is

$$\left[i\frac{\Box}{\xi}\partial_\lambda \frac{\delta}{\delta J_\lambda(x_1)} - \partial_\lambda J^\lambda(x_1) + e\bar{\zeta}\frac{\delta}{\delta \bar{\zeta}(x_1)} - e\zeta\frac{\delta}{\delta \zeta(x_1)}\right]Z[\zeta, \bar{\zeta}, J] = 0. \quad (4.50)$$

The reason that the integral which was present in 4.48 is absent in 4.49 is because the integral must vanish for arbitrary Λ, (for example we could think of choosing $\Lambda(y) = \varepsilon\delta(y - x_1)$ where ε is small and x_1 is fixed). The equation for $Z[\zeta, \bar{\zeta}, J]$ can be transformed into an equation for $\Gamma[\bar{\psi}, \psi, A]$ the generator of proper vertices. This we shall now do. Firstly we obtain the equation for $W[\zeta, \bar{\zeta}, A]$ that results from using the relation $Z = \exp[iW]$. This equation is almost the same as 4.50

$$\left[i\frac{\Box}{\xi}\partial_\lambda \frac{\delta}{\delta J_\lambda(x_1)} - \partial^\lambda J_\lambda(x_1)W^{-1} + e\bar{\zeta}\frac{\delta}{\delta\bar{\zeta}(x_1)} - e\zeta\frac{\delta}{\delta\zeta(x_1)}\right]W[\zeta, \bar{\zeta}, J] = 0. \quad (4.51)$$

Then if we use the definition of Γ given in 1.234 of Chapter 1 we obtain the partial functional differential equation that we require. It is

$$i\frac{\Box}{\xi}\partial^\lambda A_\lambda(x_1) + \partial^\lambda \frac{\delta\Gamma}{\delta A_\lambda(x_1)} - e\psi\frac{\delta\Gamma}{\delta\psi(x_1)} + e\bar{\psi}\frac{\delta\Gamma}{\delta\bar{\psi}(x_1)} = 0. \quad (4.52)$$

where we have made liberal use of the following relations which follow from defining Γ as the functional Legendre transform of W namely

$$\frac{\delta W}{\delta J_\lambda(x)} = A_\lambda(x) \qquad \frac{\delta \Gamma}{\delta A_\lambda(x)} = -J_\lambda(x)$$

$$\frac{\delta W}{\delta \zeta(x)} = \bar{\psi}(x) \qquad \frac{\delta \Gamma}{\delta \bar{\psi}(x)} = -\zeta(x) \qquad (4.53)$$

$$\frac{\delta W}{\delta \bar{\zeta}(x)} = \psi(x) \qquad \frac{\delta \Gamma}{\delta \psi(x)} = -\bar{\zeta}(x).$$

It is frequently deemed convenient to eliminate the first term in 4.52 by defining the generating function Γ_0 where

$$\Gamma_0 = \Gamma - \int \frac{dx}{2\xi} (\partial_\lambda A^\lambda)^2. \qquad (4.54)$$

Then we find the equation for Γ_0

$$\partial^\lambda \frac{\delta \Gamma_0}{\delta A_\lambda(x_1)} - e\psi \frac{\delta \Gamma_0}{\delta \psi(x_1)} + e\bar{\psi} \frac{\delta \Gamma_0}{\delta \bar{\psi}(x_1)} = 0; \qquad (4.55)$$

we shall not need to use this slightly more compact form. The reason that it can be useful is that eq. 4.55 is simply the statement in differential form of the gauge invariance of the functional $\Gamma_0[\bar{\psi}, \psi, A]$. That is to say eq. 4.55 follows straightaway from requiring

$$\Gamma_0[\bar{\psi} - ie\Lambda\bar{\psi}, \psi + ie\Lambda\psi, A_\mu + \partial_\mu \Lambda] = \Gamma_0[\bar{\psi}, \psi, A_\mu] \qquad (4.56)$$

and then expanding the LHS in a Taylor series to first order in Λ and setting the coefficient of Λ to zero. Equation 4.52 is our basic relation. Our first use of it is to derive the Ward identity used in Chapter 3. To do this we functionally differentiate 4.52 with respect to $\psi(x_2)$, $\bar{\psi}(x_3)$ and set $A_\lambda = \psi = \bar{\psi} = 0$. We obtain the relation

$$\partial^\lambda_{x_1} \frac{\delta^3 \Gamma[0]}{\delta \bar{\psi}(x_3)\, \delta \psi(x_2)\, \delta A_\lambda(x_1)} = e\delta(x_1 - x_2) \frac{\delta^2 \Gamma[0]}{\delta \bar{\psi}(x_3)\, \delta \psi(x_1)}$$

$$- e\delta(x_1 - x_3) \frac{\delta^2 \Gamma[0]}{\delta \bar{\psi}(x_1)\, \delta \psi(x_2)}.$$

$$(4.57)$$

We want to make use of this relation in momentum space. We therefore Fourier-transform eq. 4.57. The variables conjugate to x_1, x_2, x_3

are q, p, and r respectively. On performing the Fourier transform an energy-momentum conserving delta function occurs on both sides of the equation after deleting this we find

$$q^\mu \Gamma_\mu(p, q, p+q) = S_F^{-1}(p+q) - S_F^{-1}(p) \tag{4.58}$$

where we are using the notation of Chapter 3. But this is precisely the Ward identity 3.10 of Chapter 3 from which we derived $Z_1 = Z_2$. We obviously have in our hands a powerful tool for deriving relations between Green's functions that follow from gauge invariance. We shall call all such relations Ward identities. It is worth seeing what else we can derive from 4.52. For example if we differentiate with respect to $A_\nu(x_2)$ and set $A = 0$ we obtain

$$\partial^\mu \frac{\delta^2 \Gamma[0]}{\delta A_\mu(x_1)\, \delta A_\nu(x_2)} = \frac{i\Box}{\xi} \partial^\nu \delta(x_1 - x_2). \tag{4.59}$$

Now $\delta^2 \Gamma[0]/\delta A_\mu(x_1)\, \delta A_\nu(x_2)$ is minus the inverse propagator as we showed in Chapter 1. If we translate the content of 4.59 into momentum space we obtain

$$q^\mu \Gamma_{\mu\nu}(q) = q_\nu \frac{q^2}{\xi} \tag{4.60}$$

where $\Gamma_{\mu\nu}(q)$ is the inverse propagator in momentum space which is defined by ($D_{\mu\nu}(q)$ is the photon propagator)

$$\Gamma_{\mu\lambda}(q) D_{\lambda\nu}(q) = g_{\mu\nu}. \tag{4.61}$$

But $\Gamma_{\mu\nu}(q)$ is a symmetric second rank tensor and can only depend on $g_{\mu\nu}$ and $q_\mu q_\nu$, i.e.

$$\Gamma_{\mu\nu} = d^{-1}(q^2)_{\mu\nu} q^2 + e(q^2) q_\mu q_\nu \tag{4.62}$$

where d^{-1}† and e are two functions of q^2 and the q^2 is inserted to keep them both dimensionless. The requirement 4.60 immediately shows that

$$\Gamma_{\mu\nu} = d^{-1}(q_{\mu\nu} q^2 - q_\mu q_\nu) + \xi^{-1} q_\mu q_\nu. \tag{4.63}$$

Inverting this equation using 4.61 gives the photon propagator in the form

$$D_{\mu\nu}(q) = \frac{d}{q^2}\left(q_{\mu\nu} - \frac{q_\mu q_\nu}{q^2}\right) + \xi \frac{q_\mu q_\nu}{q^4}. \tag{4.64}$$

† The notation d^{-1} is chosen to conform with conventional practice.

This result does not at first sight appear very interesting but in fact it is a completely general form for the photon propagator. Moreover it was derived without calculating any Feynman diagrams, we used only a Ward identity (of course, the unknown function d must be calculated by Feynman diagrams). A point which is well worth making that the purely longitudinal part of the photon propagator, i.e. the coefficient of ξ remains $q_\mu q_\nu / q^4$ as it was for the free propagator. All the details of the interaction are contained in $d(q^2)$ the coefficient of the transverse part of the propagator.

A further example of the use of eq. 4.52 is provided by differentiating with respect to A and passing to momentum space. We can then obtain the photon–photon proper vertex $\Gamma_{\lambda\mu\nu\sigma}(p, q, r, s)$, $p+q+r+s = 0$. The Ward identity then says that

$$p^\lambda \Gamma_{\lambda\mu\nu\sigma} = 0 \qquad (4.65)$$

but since the momentum p_λ could be any of the four possible momenta we also have

$$q^\mu \Gamma_{\lambda\mu\nu\sigma} = r^\nu \Gamma_{\lambda\mu\nu\sigma} = s^\sigma \Gamma_{\lambda\mu\nu\sigma} = 0, \qquad (4.66)$$

The relations 4.65, 66 greatly simplify the tensor structure of the photon–photon scattering amplitude. It is also clear that a similar property holds for the scattering of $2n$ photons when $n > 2$. The reader can himself construct Ward identities which involve Fermions and are distinct from 4.58. In concluding this section it must be said that gauge invariance is a powerful tool for simplifying and elucidating the structure of many different Green's functions.

An Instructive Example

In this section we give a concrete example of how a completely different gauge fixing term to $\int dx \, \{-\frac{1}{2}(\partial_\lambda A^\lambda)^2\}$ can still give rise to the same S-matrix elements.[2] The gauge-fixing term $\int dx \, \{-\frac{1}{2}(\partial_\lambda A^\lambda)^2\}$ is replaced by

$$H = \int dx \, \{-\tfrac{1}{2}(\partial_\lambda A^\lambda + g A_\mu A^\mu)^2\}. \qquad (4.67)$$

We note that H is not invariant under gauge transformations as it should not be. Further we observe that it is no longer only a quadratic in A_μ as the previous gauge-fixing term was, it contains cubic and quartic terms in A_μ. These terms will cause us to expand the Feynman rules for QED so as to include trilinear and quadrilinear vertices for photons. There is also the question of the form of the ghost term $\exp[iG]$. According to 4.9,10 the ghost term is given by the equation

$$\exp[-iG] = \int \mathscr{D}[\Lambda] \exp\left[i\int dx \left\{-\tfrac{1}{2}(\partial_\lambda A^\lambda_\Lambda + gA^2_\Lambda)^2\right\}\right]. \quad (4.68)$$

If we expand $\partial_\lambda A^\lambda_\Lambda + gA^2_\Lambda$ to the first order in Λ we obtain

$$\exp[-iG] = \int \mathscr{D}[\Lambda] \exp\left[i\int dx\right.$$
$$\left. \cdot \left\{-\tfrac{1}{2}(\partial_\lambda A^\lambda + gA^2_\Lambda + \Box\Lambda + 2g\partial_\lambda \Lambda A^\lambda + O(\Lambda^2))^2\right\}\right]. \quad (4.69)$$

With the techniques of Chapter 1 it is a simple matter to change variables and write 4.69 as

$$\exp[-iG] = \int \mathscr{D}[c]\mathscr{D}[\bar{c}] \exp\left[i\int dx \left\{-(\partial_\mu \bar{c})(\partial^\mu c) + 2g\bar{c}A_\mu \partial^\mu c\right\}\right]. \quad (4.70)$$

Where $\bar{c}$ is the complex conjugate of c; the appearance of a complex field to integrate over results from the fact that the photon interacts with charged fields. The total addition to the usual QED Lagrangian is therefore

$$-\tfrac{1}{2}(\partial_\mu A^\lambda + g^2 A^2_\lambda)^2 - (\partial_\mu \bar{c})(\partial^\mu c) + 2g\bar{c}A_\mu \partial^\mu c. \quad (4.71)$$

The total result of this peculiar treatment of the gauge problem can be read off from eq. 4.71. The complex field c interacts with the photon as would a charged scalar field apart from its having the opposite sign of a conventional charged scalar field. Therefore it is a ghost field with Fermi-statistics and each closed loop of ghosts receives a minus sign. There are also trilinear and quadrilinear vertices for the coupling of photons. The Feynman rules for these additional interactions are

ghost propagator ----------●---------- $\dfrac{-i}{p^2}$

ghost vertex $-2gp_\mu$

GAUGE AND INFRARED PROPERTIES

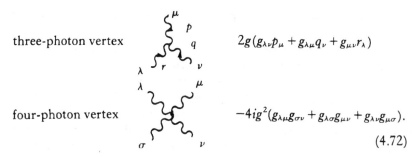

three-photon vertex $\quad 2g(g_{\lambda\nu}p_\mu + g_{\lambda\mu}q_\nu + g_{\mu\nu}r_\lambda)$

four-photon vertex $\quad -4ig^2(g_{\lambda\mu}g_{\sigma\nu} + g_{\lambda\sigma}g_{\mu\nu} + g_{\lambda\nu}g_{\mu\sigma}).$

(4.72)

With all these changes it is difficult to believe that QED has not been completely destroyed. However, when one performs calculations one finds that the diagrams which result from these new Feynman rules all cancel out. For example we can calculate the lowest order contributions in g and e to the photon propagator. The diagrams are shown in

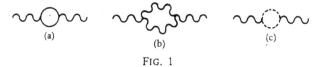

FIG. 1

Fig. 1 (a), (b), (c). The integration terms in 4.71 have, of course, been normal ordered. In Fig. 1(b) there must be a combinatorial factor of $\frac{1}{2}$. In Fig. 1(c) the closed ghost loop receives a minus sign for Fermi statistics. The total contribution to the photon propagator is obtained by adding together all three diagrams. Miraculously diagrams (b) and (c) exactly cancel leaving the usual Fermion loop diagram of Fig. 1(a). Similar calculations have the same property: dependence on the fictitious coupling constant g cancels out. It is usual to associate such complications as ghosts and ghost interactions with non-Abelian gauge theories.[1] As we have seen here though QED can exhibit all these features if the gauge is chosen appropriately. The difference between the two sorts of theories lies in the fact that for QED there is a simple gauge which is satisfactory in every way and ghosts never appear. But in non-Abelian gauge theories there is no simple satisfactory gauge without interacting ghosts.

Ward Identity and Unitarity

We have given some of the uses of the Ward identity for various Green's functions in QED. One of the principal results that we are able to prove using the gauge properties of QED is $Z_1 = Z_2$. This result obviously shortens at a stroke the calculations necessary to renormalize QED. This fact is quite rightly highlighted in most discussions on gauge invariance and the Ward identities. We would now like to emphasize that the proving of unitarity for QED via the Ward identities is of equal status and importance to this. Unfortunately QED formulated in a manifestly Lorentz covariant fashion is not manifestly unitary. The converse is also true. Of course, QED is both a Lorentz invariant and a unitary theory but we have to do extra work to prove this. Let us see why there is not manifest unitarity. We choose to work in the Feynman gauge which corresponds to $\xi = 1$. The free photon propagator in configuration space is then given by the equation

$$\langle 0|T(A_\mu(x)A_\nu(0))|0\rangle = \frac{-1}{2\pi^2}\frac{g_{\mu\nu}}{x^2 - i\varepsilon} \tag{4.73}$$

while the VEV of the product of two fields $A_\mu(x)$ and $A_\nu(0)$ is obtained by replacing $i\varepsilon$ by $i\varepsilon x_0$ where

$$\langle 0|A_\mu(x)A_\nu(0)|0\rangle = \frac{-1}{2\pi^2}\frac{g_{\mu\nu}}{x^2 - i\varepsilon x_0}. \tag{4.74}$$

(One can verify that 4.74 gives rise to 4.73.) We now insert a complete set of states and use translation invariance, i.e. we use

$$1 = \sum_H \int \mathrm{d}p_H \, |H\rangle\langle H|$$

$$A_\mu(x) = \mathrm{e}^{\mathrm{i}Px} A_\mu(0) \, \mathrm{e}^{-\mathrm{i}Px} \tag{4.75}$$

where p_H is the momentum of the state labelled H and P_μ is the energy momentum operator. We can now rewrite 4.74 as

$$\int \mathrm{d}p_H \, \mathrm{e}^{-\mathrm{i}p_H x} \langle 0|A_\mu(0)|H\rangle\langle H|A_\nu(0)|0\rangle = \frac{-g_{\mu\nu}}{2\pi^2(x^2 - i\varepsilon x_0)}. \tag{4.76}$$

Now define the spectral function $\mathcal{P}_{\mu\nu}(p_H)$ by

$$\mathcal{P}_{\mu\nu}(p_H) = (2\pi)^3 \langle 0|A_\mu(0)|H\rangle\langle H|A_\nu(0)|0\rangle. \tag{4.77}$$

But because all physical states should have positive energy and cannot have spacelike four momenta we have the identity

$$\mathcal{P}_{\mu\nu}(p_H) = \theta(p_0^H)\theta(p_H^2)\mathcal{P}_{\mu\nu}(p_H^2). \tag{4.78}$$

Using the further identity

$$\int_0^\infty d\sigma^2 \, \mathcal{P}_{\mu\nu}(\sigma^2) \, \delta(\sigma^2 - p_H^2) = \mathcal{P}_{\mu\nu}(p_H^2)\theta(p_H^2) \qquad (4.79)$$

we derive the representation

$$\frac{-g_{\mu\nu}}{2\pi^2(x^2 - i\varepsilon x_0)} = \frac{1}{(2\pi)^3} \int_0^\infty d\sigma^2 \int d p_H \, \theta(p_0^H) \, \delta(\sigma^2 - p_H^2) \mathcal{P}_{\mu\nu}(\sigma^2) \, e^{-ip_H x} \qquad (4.80)$$

The integral over p_H is a standard one so that we obtain

$$\frac{-g_{\mu\nu}}{2\pi^2(x^2 - i\varepsilon x_0)} = -i \int_0^\infty d\sigma^2 \, \mathcal{P}_{\mu\nu}(\sigma^2) \, \Delta^-(x, \sigma^2) \qquad (4.81)$$

where

$$\Delta^-(x, \sigma^2) = \frac{i}{(2\pi)^3} \int dk \, e^{-ikx} \theta(k_0) \, \delta(k^2 - \sigma^2). \qquad (4.82)$$

Equation 4.81 is the standard spectral representation for the two-point function. We wish to make use of it in momentum space. The Fourier transform can be easily done using 4.82; we thus obtain

$$\frac{-g_{\mu\nu}}{2\pi^2} \int dx \frac{e^{ipx}}{x^2 - i\varepsilon x_0} = \frac{1}{(2\pi)^3} \int_0^\infty d\sigma^2 \, \mathcal{P}_{\mu\nu}(\sigma^2) \theta(p_0) \, \delta(p^2 - \sigma^2). \qquad (4.83)$$

The spectral function $\mathcal{P}_{\mu\nu}(p^2)$ can now be seen to be diagonal in μ and ν. Each of the diagonal elements has the interpretation of being a probability. For example

$$\mathcal{P}_{00} = (2\pi)^3 \langle 0|A_0(0)|H\rangle\langle H|A_0(0)|0\rangle$$
$$= (2\pi)^3 |\langle 0|A_0(0)|H\rangle|^2 \qquad (4.84)$$

and

$$\mathcal{P}_{33} = (2\pi)^3 \langle 0|A_3(0)|H\rangle\langle H|A_3(0)|0\rangle$$
$$= (2\pi)^3 |\langle 0|A_3(0)|H\rangle|^2. \qquad (4.85)$$

However, returning to 4.83 we see that since $g_{00} = -g_{33}$ it is not possible that both $\mathcal{P}_{00}$ and $\mathcal{P}_{33}$ are positive. One of $\mathcal{P}_{00}$ and $\mathcal{P}_{33}$ must therefore be a negative rather than a positive probability. It is easy to check from 4.83 that $\mathcal{P}_{00}$ is negative and $\mathcal{P}_{11}$, $\mathcal{P}_{22}$, and $\mathcal{P}_{33}$ are positive.

The occurrence of negative probability states created by the time-like photon A_0 calls unitarity into question, for the space of states no longer has a positive metric. This does not happen. The reason that this difficulty has arisen is because there are only two independent photon degrees of freedom. The spectral representation treats all four components of A_μ equally and the occurrence of negative probability is a sign that all these four components cannot be alike in every respect. We recall[3] that physical photons are distinguished by the fact that they are transverse; they have polarization vectors which satisfy:

$$\varepsilon_\mu(k, \lambda)k^\mu = 0, \qquad k^2 = 0 \qquad \lambda = 1, 2 \tag{4.86}$$

The momentum of the photon is k_μ, and λ labels the two possible transverse states of polarization. There are also two other polarization vectors $\varepsilon_\mu(k, \lambda)$, $\lambda = 0, 3$ and the four together satisfy, (in the Feynman gauge where $\xi = 1$).

$$\sum_{\lambda=0}^{3} \varepsilon_\mu(k, \lambda)\varepsilon_\nu(k, \lambda) = -g_{\mu\nu}. \tag{4.87}$$

We shall show that QED is unitary in a space spanned by only the two transverse polarization states of 4.86. The two other states decouple from S-matrix elements. Before proving this we need some preliminary properties of the polarization sums. If we sum over only the transverse degrees of freedom we obtain

$$\sum_{\lambda=1,2} \varepsilon_\mu(k, \lambda)\varepsilon_\nu(k, \lambda) = -g_{\mu\nu} + a_{\mu\nu} \tag{4.88}$$

where 4.88 serves to define the tensor $a_{\mu\nu}$. But in view of 4.87 we must also have

$$\sum_{\lambda=0,3} \varepsilon_\mu(k, \lambda)\varepsilon_\nu(k, \lambda) = -a_{\mu\nu}. \tag{4.89}$$

The tensor $a_{\mu\nu}$ is symmetric in μ and ν by its definition. We can determine a choice for $a_{\mu\nu}$ by the transversality condition 4.86. Let $\bar{k}_\mu$ be the vector k_μ with the sign of its energy component negative rather than positive, i.e.

$$\bar{k}_\mu = (-k_0, \mathbf{k}), \tag{4.90}$$

then let

$$a_{\mu\nu} = \frac{\bar{k}_\mu k_\nu + k_\mu \bar{k}_\nu}{(k \cdot \bar{k})}. \tag{4.91}$$

This definition satisfies the transversality condition 4.86 so that in summary we have

$$\sum_{\lambda=1,2} \varepsilon_\mu(k,\lambda)\varepsilon_\nu(k,\lambda) = -g_{\mu\nu} + \frac{\bar{k}_\mu k_\nu + k_\mu \bar{k}_\nu}{(k \cdot \bar{k})} \qquad (4.92)$$

and

$$\sum_{\lambda=0,3} \varepsilon_\mu(k,\lambda)\varepsilon_\nu(k,\lambda) = \frac{-(\bar{k}_\mu k_\nu + k_\mu \bar{k}_\nu)}{(k \cdot \bar{k})}. \qquad (4.93)$$

We can now verify unitarity. We choose as a typical S-matrix element the photon–photon scattering amplitude. Since we always deal with connected S-matrix elements we replace the unitarity condition

$$SS^\dagger = I \qquad (4.94)$$

by

$$-i(T - T^\dagger) = TT^\dagger \qquad (4.95)$$

where the S-matrix S is related to T by the equation

$$S = I + iT. \qquad (4.96)$$

We can represent the unitarity equation (4.96) diagrammatically as shown in Fig. 2.[3] Figure 2 shows the unitarity equation for the

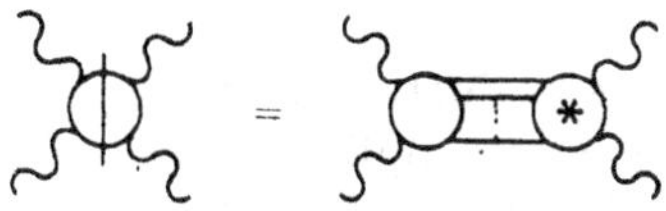

FIG. 2

connected photon–photon scattering amplitude. It is obtained by taking matrix elements between two-photon states of eq. 4.95. We restrict the momenta to give forward scattering so that Fig. 2 represents the optical theorem for the total photon–photon cross-section. The notation of Fig. 2 needs some explanation. The vertical line on the LHS of the equation denotes taking the imaginary part of the amplitude. An asterisk denotes the complex conjugate of an amplitude. The internal lines on the RHS of the equation denote insertion of a complete set of intermediate states[†]. Suppose then that

---

[†] The complete set of intermediate states are by definition made up of all those states that are physical, i.e. electrons and transverse photons.

the external lines are physical, i.e. transverse photons. Then we must show that only transverse photons contribute to the unitarity equation of Fig. 2. Let us take a typical Feynman graph contributing to the LHS of Fig. 2. This graph is displayed in Fig. 3. In perturbation

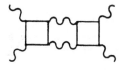

FIG. 3

theory contributions to the imaginary part are obtained by 'cutting' the diagram with the vertical line in all possible ways.[4] For example some cut diagrams are shown in Fig. 4. Because we only have misgivings about photon propagators we shall not be concerned with Fig. 4(a) where only electron lines are cut. For definiteness we choose to

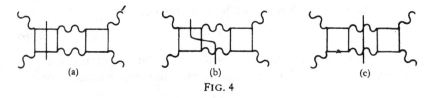

FIG. 4

deal with Fig. 4(c). The evaluation of the imaginary part denoted by the cutting procedure is accomplished[4] by replacing each cut propagator $-g_{\mu\nu}/q^2$ by $g_{\mu\nu}\theta(q_0)\,\delta(q^2)(2\pi)$. To make this procedure clear for Fig. 4(c) we break up the diagram into two halves and label all the momenta. With the definitions given in Fig. 5† the Feynman integral

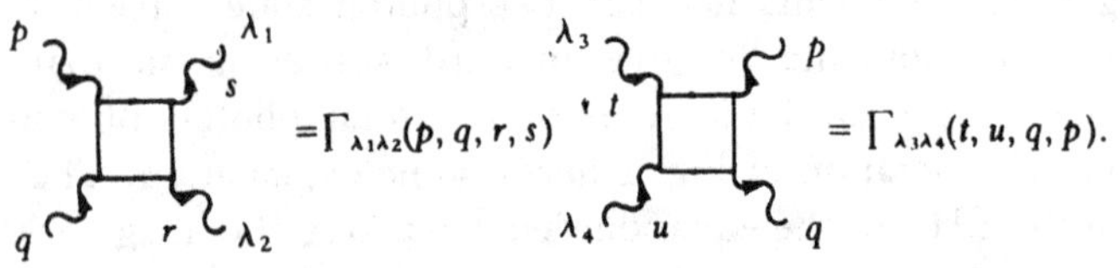

$$p \searrow \quad \overset{\lambda_1}{\curvearrowright} \quad s \qquad \overset{\lambda_3}{\curvearrowright} \quad \nearrow p$$
$$\boxed{} = \Gamma_{\lambda_1\lambda_2}(p,q,r,s) \quad {}^{,t}\boxed{} = \Gamma_{\lambda_3\lambda_4}(t,u,q,p).$$
$$q \nearrow \quad r \quad \lambda_2 \qquad \lambda_4 \quad u \quad \searrow q$$

FIG. 5

† The quantities $\Gamma_{\lambda_1\lambda_2}$ and $\Gamma_{\lambda_3\lambda_4}$ have only two tensorial indices rather than four because the external lines p and q already are contracted with polarization vectors as is appropriate for an S-matrix element.

of Fig. 4(c) is simply A where A is

$$-\int \frac{dk}{(2\pi)^4}\Gamma_{\lambda_1\lambda_2}(p, q, k, p+q-k)\frac{g^{\lambda_1\lambda_3}}{(k-p-q)^2}\frac{g^{\lambda_2\lambda_4}}{k^2}$$
$$\cdot \Gamma_{\lambda_3\lambda_4}(p+q-k, k, q, p) \qquad (4.97)$$

therefore the imaginary part of A is given by the integral

$$-\int \frac{dk}{(2\pi)^4}\Gamma_{\lambda_1\lambda_2}g^{\lambda_1\lambda_2}g^{\lambda_3\lambda_4}\theta(k_0)\,\delta(k^2)\theta(k_0-p_0-q_0)\delta\{(k-p-q)^2\}\Gamma_{\lambda_3\lambda_4}. \qquad (4.98)$$

Because of the presence of the δ functions the integral over k can be easily done and 4.98 becomes:

$$-\int \frac{d\Omega}{(2\pi)^2}\Gamma_{\lambda_1\lambda_2}(p, q, \hat{k}, \hat{p}+\hat{q}-\hat{k})g^{\lambda_1\lambda_3}g^{\lambda_2\lambda_4}\Gamma_{\lambda_3\lambda_4}(\hat{p}+\hat{q}-\hat{k}, \hat{k}, q, p). \qquad (4.99)$$

In 4.99 $\hat{k}$ and $\hat{p}+\hat{q}-\hat{k}$ are on-shell momenta; i.e.

$$\hat{k}^2 = 0, \qquad (\hat{p}+\hat{q}-\hat{k})^2 = 0 \qquad (4.100)$$

the integral over Ω is just the integral over the solid-angle in the centre of mass of the two photons. We have now produced an expression which only involves physical (i.e. on-mass shell) momenta and we can use some of the properties of transverse photons. Equation 4.87 allows us to replace the tensors $g^{\lambda_1\lambda_3}$ and $g^{\lambda_2\lambda_4}$ by sums over polarization vectors so that 4.99 becomes

$$-\int \frac{d\Omega}{(2\pi)^2}\Gamma^{\lambda_1\lambda_2}\sum_{\alpha=0}^{3}\varepsilon_{\lambda_1}(\hat{p}+\hat{q}-\hat{k},\alpha)\varepsilon_{\lambda_3}(\hat{p}+\hat{q}-\hat{k},\alpha)$$
$$\cdot \sum_{\beta=0}^{3}\varepsilon_{\lambda_2}(\hat{k},\beta)\varepsilon_{\lambda_4}(\hat{k},\beta)\Gamma^{\lambda_3\lambda_4}. \qquad (4.101)$$

Now if we consider the sum over the longitudinal polarizations we know already that

$$\sum_{\alpha=0,3}\varepsilon_{\lambda_1}(\hat{p}+\hat{q}-\hat{k},\alpha)\varepsilon_{\lambda_3}(\hat{p}+\hat{q}-\hat{k},\alpha)$$
$$= -\frac{\{(\overline{\hat{p}+\hat{q}-\hat{k}})_{\lambda_1}(\hat{p}+\hat{q}-\hat{k})_{\lambda_3} + (\hat{p}+\hat{q}-\hat{k})_{\lambda_1}(\overline{\hat{p}+\hat{q}-\hat{k}})_{\lambda_2}\}}{(\hat{p}+\hat{q}-\hat{k})\cdot(\overline{\hat{p}+\hat{q}-\hat{k}})}$$
$$\sum_{\beta=0,3}\varepsilon_{\lambda_2}(\hat{k},\beta)\varepsilon_{\lambda_4}(\hat{k},\beta) = -\frac{(\bar{\hat{k}}_{\lambda_2}\hat{k}_{\lambda_4} + \hat{k}_{\lambda_2}\bar{\hat{k}}_{\lambda_4})}{(\hat{k}\cdot\bar{\hat{k}})}. \qquad (4.102)$$

But using the Ward identity for photon–photon scattering derived in eqs 4.65, 66 we conclude that

$$(\hat{p} + \hat{q} - \hat{k})_{\lambda_1} \Gamma^{\lambda_1 \lambda_2} = 0, \qquad \hat{k}_{\lambda_2} \Gamma^{\lambda_1 \lambda_2} = 0 \qquad (4.103)$$

and other similar relations. This means that because of the Ward identities 4.103 the sum over longitudinally polarized photon states contributes precisely zero to 4.101. The sum over the transversely polarized states, of course, is not zero. Therefore we see that although states other than transversely polarized states propagate along the internal lines only the transversely polarized states survive when checking the unitarity equations for S-matrix elements. As far as the optical theorem represented in Fig. 2 is concerned only transverse photons contribute to both the LHS and RHS of the equation. Needless to say the transverse polarization states are contracted into $\Gamma^{\lambda_1 \lambda_2}$ and $\Gamma^{\lambda_1 \lambda_4}$ so that the whole expression 4.101 is one involving S-matrix elements.

Therefore we may summarize the state of affairs by saying that the restriction of the scattering matrix S to the space of transversely polarized photons provides us with a unitary operator. In the language of probability the negative probability state has cancelled against one of positive probability leaving the two transverse states which have positive probability. The purpose of this section was to verify the unitarity of QED. An important lesson that we learned on the way was that gauge invariance is essential to the proof. This result must be regarded as being of no less importance than the result $Z_1 = Z_2$ which also follows from gauge invariance.

Infrared Divergences of QED

The presence of zero-mass particles in a quantum field theory can give rise to special problems in some theories. QED is an example, non-Abelian gauge theories being another. The zero-mass particle in QED is, of course, the photon. In fact there are what are called infrared divergences in QED. These are quite distinct and different in their origin to the ultraviolet divergences that we have already encountered. Ultraviolet divergences would be present even in a version of QED where the photon is massive. Infrared divergences would disappear if the photon were massive. The infrared problems of QED are often played down a little or even dismissed in comparison with the problems created by ultraviolet divergences. Nevertheless the literature contains many papers on the infrared problems.[4] It would be fair to say that the infrared problems of QED are not at all trivial even though they may be solved in a much more satisfactory manner than the ultraviolet problems. The fact is, as we shall see, that infrared divergences cancel from physical measured quantities. Furthermore this cancellation occurs without any special techniques like infinite counter terms being necessary. Before reaching the stage where this cancellation will be clear to us we need some introduction to the sort of problem we have at hand.

We have already encountered two infrared problems in Chapter 3. These were divergences which were uncovered when renormalizing the electron propagator and the electron–photon proper vertex. We recall first the problem of the electron propagator $S_F(p)$. A divergence arose when we assumed that

$$\lim_{p^2 \to m^2} S_F(p) \to \frac{Z_2}{\not{p} - m} + S(p) \qquad (4.104)$$

with $S(p)$ a regular function of p. Instead of 4.104 holding we found that there was a logarithmic singularity $\ln(1 - p^2/m^2)$ in $S_F(p)$ as well as the simple pole displayed in 4.104. This contradicts the assumption that $S(p)$ is a regular function since it has a branch cut at $p^2 = m^2$. At the time we were constructing a suitable definition for the renormalization constant Z_2 and so we chose a different point in momentum space at which to define Z_2. The point $p_\mu = 0$ was chosen as this avoids the infrared problem since the logarithm is regular at $p_\mu = 0$. Nevertheless the logarithm is present in $S_F(p)$ and as was shown in Chapter 3 the branch cut in the complex p^2 plane beginning at $p^2 = m^2$ has a physical significance. The branch cut is present because there is a threshold for producing photons which begins at $p^2 = m^2$. Now we see a problem, how many photons can be present if $p^2 > m^2$? The answer is infinitely many! Consider the physics of the problem in the following way: let ΔE be the inaccuracy in measuring the energy of a single electron. Then this energy ΔE may be made of photons. (It is possible experimentally to achieve $\Delta E < 2m$ so that there are no positron electron pairs contributing to ΔE). If $E_1, E_2, \ldots$ are an infinite set of photons with four momenta $p_1, p_2, \ldots$ given by

$$p_i = (E_i, 0, 0, E_i)$$
$$p_i^2 = 0, \qquad i = 1, 2, \qquad (4.105)$$

then choose the nth photon E_n to have energy $\lambda\,\Delta E(6/\pi^2)(1/n^2)$ where λ is a positive real number less than unity. The total energy of this infinite cloud of photons is simply

$$\frac{6}{\pi^2}\lambda\,\Delta E \sum_1^\infty \frac{1}{n^2} = \lambda\,\Delta E < \Delta E. \qquad (4.106)$$

Thus we see that because the lower bound on the energy of a photon is zero rather than a positive number, an infinite number of photons may have an arbitrarily small total energy. We cannot therefore ever conclude that we have a single electron with uncertainty in energy ΔE. We must always allow for an infinite number of accompanying photons. This means that QED has finite energy states with infinite occupation number. The ordinary Fock space of asymptotic states is therefore inadequate for constructing the space of states for QED since it does not contain such states. Linear spaces can be constructed which contain these states.[5] We shall not describe these spaces. But we shall introduce the formalism of coherent states which allows one to compute states of infinite occupation number without direct reference to the space to which they belong. Returning to the electron propagator we note that the infrared singularity is a "mass-shell singularity", i.e. one that is discovered when the momentum of the charged line approaches its mass-shell value. This is also true of the infrared singularity in the vertex. The form factor $F(q^2)$ was found to have a logarithmic singularity, $\ln(1-p^2/m^2)$, just as was the case for the electron propagator. This is a general feature of infrared divergences. In summary, infrared divergences appear in the Green's functions of QED when the momenta of external charged lines approach their mass shell values. It seems clear that two things have to be done to deal with the infrared divergences of QED: one is to deal with the infinite occupation number of states and the other is to isolate and examine the infrared divergences of the Green's functions. We shall deal with the question of infinite numbers of photons first.

Coherent States

We have said that the Fock space of asymptotic states needs to be enlarged when considering 'soft' photons (soft photons are photons with very small energies such as those contributing to the sum in 4.106). We require states which contain infinite numbers of photons;

their occupation number is infinity, we call these coherent states. Broadly speaking we wish to replace the normal Fock space of asymptotic 'in' and 'out' states by a space of coherent states. Coherent states are quite simple to construct. They are in essence made up of an infinite sum of states with ever-increasing numbers of photons. This means that they can be defined in terms of the original Fock space of in or out states. We need to return to the usual Fock space to establish some facts and some notation that we shall make use of. First of all we need the creation operators of the radiation field of momentum k and polarization λ. We continue to work in the Feynman gauge. The creation operator is written $a^\dagger(k, \lambda)$ where

$$a^\dagger(k, \lambda) = \varepsilon^\mu(k, \lambda) a^\dagger_\mu(k) \tag{4.107}$$

$\varepsilon_\mu(k, \lambda)$ is the usual polarization vector and $a^\dagger_\mu(k)$ satisfies the commutation relation

$$[a_\mu(k), a^\dagger_\nu(k')] = -g_{\mu\nu} \delta(\mathbf{k} - \mathbf{k}'). \tag{4.108}$$

Then $a^\dagger(k, \lambda)$ satisfies the commutation relation

$$[a(k, \lambda), a^\dagger(k', \lambda')] = \delta(\mathbf{k} - \mathbf{k}') \delta_{\lambda\lambda'}. \tag{4.109}$$

The n-particle states of the photon's Fock space are constructed by applying powers of $a^\dagger(k, \lambda)$ to the vacuum state. For example we write

$$|n\rangle = |(k, \lambda_1) \ldots (k_n \lambda_n)\rangle = \frac{1}{(n!)^{1/2}} a^\dagger(k_1, \lambda_1) \ldots a^\dagger(k_n, \lambda_n) |0\rangle. \tag{4.110}$$

This is straightforward quantum mechanics. Next we require the form of the S-matrix in terms of $a^\dagger(k, \lambda)$. Of course we cannot in general find the form of the matrix in terms of $a^\dagger(k, \lambda)$. As is usual in physical problems we tackle this sort of difficulty by making a physically motivated approximation. The approximation that we make is that we neglect the electron term in the QED Lagrangian. That is to say we take as our Lagrangian the expression

$$-\tfrac{1}{4} F_{\mu\nu} F^{\mu\nu} - \tfrac{1}{2} (\partial_\lambda A^\lambda)^2. \tag{4.111}$$

To the Lagrangian given in 4.111 we add the source term $J_\mu A^\mu$ so that our resultant Lagrangian is L, where

$$L = -\tfrac{1}{4} F_{\mu\nu} F^{\mu\nu} - \tfrac{1}{2} (\partial_\lambda A^\lambda)^2 + J_\mu A^\mu. \tag{4.112}$$

There is a physical reason which gives us confidence that the approximation of neglecting the electron terms is a sensible approximation. This reason is that we are interested in the physical effects of very low energy or low frequency photons. In these circumstances the energy E of each photon is so small that $E \ll 2m$, (m is the electron mass); therefore these photons cannot produce electrons via polarization of the vacuum and so appear not to interact with electrons. Confronted with the Lagrangian 4.112 we recognize an expression already encountered in Chapter 1. The S-matrix which results from the Lagrangian of 4.112 is most easily constructed by starting with the functional $Z[J]$ of Chapter 1. This functional is given in eq. 1.215 of Chapter 1 with $\zeta = 0$. We have

$$Z[J] = \exp\left[\frac{-i}{2}\int dx\,dy\,J_\mu(x)D_F^{\mu\nu}(x-y)J_\nu(y)\right]. \quad (4.113)$$

According to Chapter 1 the S-matrix elements are obtained by taking the Green's functions which contribute to $Z[J]$ after differentiating with respect to J. In coordinate space we write such a Green's function as $\Gamma_{\mu_1\ldots\mu_n}(x_1,\ldots x_n)$ where Γ is defined by

$$\Gamma_{\mu_1\ldots\mu_n}(x_1,\ldots x_n) = \prod_{i=1}^n i^n \int dy\, D_{\mu_i F}^\nu(x_i - y) J_\nu(y) Z[J] \quad (4.114)$$

(note in these calculations we do not set $J = 0$ at the end, we regard J as an external field). In momentum space this relation becomes

$$\Gamma_{\mu_1\ldots\mu_n}(k_1,\ldots k_n) = Z[J]\prod_{i=1}^n \frac{\hat{J}_{\mu_i}(k_i)}{k_i^2} \quad (4.115)$$

where $\hat{J}_\mu(k)$ is the Fourier transform of $J_\mu(x)$. Now an S-matrix element is defined in terms of the in and out states by

$$\langle(k_1\lambda_1)\ldots(k_n\lambda_n)_{\text{out}}|0_{\text{in}}\rangle = \langle(k_1\lambda_1)\ldots(k_n\lambda_n)_{\text{in}}|S|0_{\text{in}}\rangle$$
$$= S[(k_1\lambda_1)\ldots(k_n\lambda_n)]. \quad (4.116)$$

The relation between $\Gamma_{\mu_1\ldots\mu_n}(k_1,\ldots k_n)$ and $S[(k_1\lambda_1)\ldots(k_n\lambda_n)]$ is immediate, it is such that $S[(k_1\lambda_1)\ldots(k_n\lambda_n)]$ is given by

$$S[(k_1\lambda_1)\ldots(k_n\lambda_n)] = (-i)^n Z[J]\prod_{i=1}^n \hat{J}_{\mu_i}(k_i)\frac{\varepsilon^{\mu_i}(k_i,\lambda_i)}{[(2\pi)^3 2k_{0i}]^{1/2}}$$
$$(4.117)$$

GAUGE AND INFRARED PROPERTIES 153

$k_i^2 = 0$. But if we use eq. 4.110 it is now easy to see that the S-matrix itself will be given by the equation

$$S = \exp\left[\frac{-i \int d^3\mathbf{k}}{[(2\pi)^3 2k_0]^{1/2}} \{\varepsilon_\mu(k,\lambda) \hat{f}^\mu(k) a^\dagger(k,\lambda) + \varepsilon_\mu(k,\lambda) \hat{f}^{*\mu}(k) a(k,\lambda)\}\right] \quad (4.118)$$

(J_μ^* is the complex conjugate of J_μ). The correctness of this form for S is easily verified by taking the matrix element defined in 4.116 and using the commutation relations 4.108, 4.109 together with eq. 4.110. After this somewhat technical preamble we are now ready to define a coherent state $|C\rangle$. The coherent state is simply defined by the equation

$$|C\rangle = S|0\rangle. \quad (4.119)$$

Since we can expand the exponential in 4.118 and produce higher and higher powers of $a^\dagger(k,\lambda)$ we see that $|C\rangle$ as defined by 4.119 contains an infinite number of photons. In fact we will hardly need to use any explicit properties of coherent states to prove that the infrared divergences cancel in QED. We do not deal with applications and computations using coherent states; a full account of these is found in reference 4, in fact the material of this section should be an aid to the reader pursuing this reference. The fact that such states are necessary is quite evident on physical grounds as we have seen above. In fact the final states of scatterings which take place among electrons and photons must contain coherent states. This is because an infinite number of on-shell photons may exist in the final state and still have only a small finite amount of the available final state energy.

Cancellation of Infrared Divergences

When approaching the problem of infrared divergences in QED we are unlikely to forget the ultraviolet divergences that are usually encountered when renormalizing the theory. We therefore ask if the divergence of a Feynman graph is due to its poor behaviour in the ultraviolet region of momenta, or its poor behaviour in the infrared region of momenta; or does it depend on both these features? There are various ways of dealing with this query. Although the details differ they all amount to much the same thing: a separation of the ultraviolet divergences from the infrared divergences. The simple method that we shall use[*] is stated as an identity for the photon propagator $D_{\mu\nu}(p)$. We write

$$D_{\mu\nu}(p) = D^s_{\mu\nu}(p) + D^h_{\mu\nu}(p) \tag{4.120}$$

where

and

$$D^s_{\mu\nu} = -\frac{ig_{\mu\nu}}{p^2} + \frac{ig_{\mu\nu}}{p^2 - M^2}$$

$$D^h_{\mu\nu} = -\frac{ig_{\mu\nu}}{p^2 - M^2}. \tag{4.121}$$

The letters s and h stand for soft and hard photons. The reason for this is easy to see. $D^s_{\mu\nu}(p)$ diverges for small p^2 and so contains the infrared divergences due to the soft photons. However for large p^2 we have

$$D^s_{\mu\nu}(p) \to \frac{iM^2}{p^4}. \tag{4.122}$$

$$p^2 \to \infty$$

Thus insertion of $D^s_{\mu\nu}(p)$ into a Feynman diagram will provide a propagator which is strongly convergent for ultraviolet momenta and produces no ultraviolet infinities. By contrast $D^h_{\mu\nu}(p)$ is finite for small p^2 but produces the usual $1/p^2$ behaviour for ultraviolet momenta. Thus $D^h_{\mu\nu}(p)$ produces ultraviolet infinities but no infrared infinities. So far as the propagator is concerned we have managed a neat separation of the infrared from the ultraviolet divergences. Still more may be accomplished along these lines. Let us consider a Green's function $G(p_1 \ldots p_n)$ and use the separation 4.120 of the photon propagator. It is then possible to prove in general that the infrared singularity in the summed Feynman graphs factorizes. That is to say we can show that[5]

$$G(p_1 \ldots p_n) = G^s(p_i) G^h(p_1 \ldots p_n) \tag{4.123}$$

where $G^h(p_1 \ldots p_n)$ is infrared finite and consists of graphs where the photon propagator $D_{\mu\nu}(p)$ is replaced by $D^h_{\mu\nu}(p)$, and $G^s(p_i)$ contains the infrared divergent part. Of course, $G^h_{\mu\nu}(p)$ will still contain ultraviolet divergences. It is fortunate then that in QED one can accomplish a complete separation of the two sources of infinities: infrared and ultraviolet. We shall not need to prove 4.123 as we shall

GAUGE AND INFRARED PROPERTIES   155

prove the infrared finiteness by a more direct method.[6] The argument that we shall present is not strictly speaking complete. The lack of completeness arises from the following considerations. We shall simply calculate the most infrared singular terms which contribute to the Green's functions and show that these singularities cancel. It might be then that some of the non-leading singularities did not cancel and therefore falsify our result. This does not in fact happen. The further details to make our treatment complete and rigorous can be found in references 4 and 5.

We attack the problem by considering the effect of inserting soft photons into an arbitrary Feynman diagram. The method consists of starting with Feynman diagrams which contain no soft photons; these are called core diagrams after Kibble.[5] More simply this means that a Feynman diagram is made into a core diagram by removing all photon lines, internal or external, which make the diagram infrared divergent. The sum of the core diagrams is the Green's function $G^{(h)}(p_1 \ldots p_n)$ of 4.123. Next we systematically add in the soft photons, calculating at every stage their contribution to the infrared divergence of the Green's function. In this way we shall eventually find that the infrared divergence cancels after all the soft photons have been added in. Let us assume then that we are dealing with the core diagrams. The soft photons that can be attached divide into two classes: external soft photons and internal soft photons. We illustrate this in Fig. 6. In Fig. 6

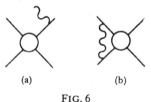

FIG. 6

we show soft photons attached to electron–electron scattering amplitude. Figure 6(a) shows the addition of an external soft photon while Fig. 6(b) shows the addition of an internal soft photon. We shall deal with the properties of external soft photons first. Let the momentum of an electron line be p and the momentum of a soft photon be q. The attaching of the soft photon to the electron line introduces an extra electron propagator of momentum $p + \zeta q$ and an extra vertex. The symbol ζ is simply ∓ 1 according to the direction of the electron line: incoming or outgoing. The extra factor that this produces in the

Feynman diagram is therefore

$$\frac{e\gamma_\mu(\not{p}+\not{q}+m)}{[(p+\zeta q)^2-m^2+i\varepsilon]}. \qquad (4.124)$$

Let ζ for the sake of definiteness be $+1$ then as q is a soft photon we only require the part of 4.129 singular as $q \to 0$. This is the expression

$$\frac{e\gamma_\mu(\not{p}+m)}{p^2+2p\cdot q-m^2+i\varepsilon}. \qquad (4.125)$$

Now let this electron line be an external line as in Fig. 6(a); this means that $p^2 = m^2$. Also let us supply the wave function of the soft photon, i.e. the factor $\varepsilon_\mu(q)/(2\pi)^{3/2}(2|\mathbf{q}|)^{1/2}$. The singular part of 4.125 is then for a general ζ

$$\zeta\frac{\varepsilon^\mu(q)p_\mu}{(p\cdot q)}. \qquad (4.126)$$

Note that we only need to consider the case where the electron line is an external one ($p^2 = m^2$). For an internal electron line p^2 is in general not equal to m^2, and as we saw in Chapter 3 the electron propagator is only infrared singular as $p^2 \to m^2$. When $p^2 \neq m^2$ the denominator $p \cdot q$ of 4.126 is $p^2+2p\cdot q-m^2$ as in 4.125 which is not singular as $q \to 0$. Returning to our analysis of external electron lines we now analyse the effect of adding two soft photons to an electron line. Let the soft photons have momenta q_1 and q_2; we may attach them to the electron in either of two orders. The resulting denominator factors are

$$\frac{1}{(p\cdot q_1)\{p\cdot(q_1+q_2)\}} + \frac{1}{(p\cdot q_2)\{p\cdot(q_2+q_1)\}}. \qquad (4.127)$$

Figure 7 shows the two orders in which the photon lines may be

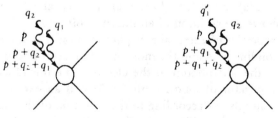

FIG. 7

attached. Conveniently we may sum the two terms in 4.127 to the expression

$$\frac{1}{(p \cdot q_1)(p \cdot q_2)}. \tag{4.128}$$

This simple piece of algebra generalizes. That is to say the effect of attaching n soft photons of momenta $q_1, \ldots q_n$ in all possible orders to an electron line of momentum p is just

$$\frac{1}{(p \cdot q_1)(p \cdot q_2) \ldots (p \cdot q_n)}. \tag{4.129}$$

If we supply the numerator factors the multiplicative factor that the core diagram receives from n external soft photons is

$$\prod_{i=1}^{n} \frac{1}{(2\pi)^{3/2}(2|\mathbf{q}_i|)^{1/2}} \zeta_i \frac{p \cdot \varepsilon(q_i)}{[p \cdot q_i]} = A. \tag{4.130}$$

There is still more to be done with this external soft photon contribution. We want eventually to find the contribution of the soft photons to a cross-section. Therefore we must take the square of the S-matrix element, sum over the photon's polarization states, and integrate over the angles at which the photons scatter with the electron, and sum n from 0 to ∞. Taking the square and summing over helicities we obtain

$$B = \frac{p^2}{n!} \prod_{i=1}^{n} \frac{-1}{(2\pi)^3 (2|\mathbf{q}_i|)} \frac{1}{(p \cdot q_i)^2}. \tag{4.131}$$

We have supplied the factor $1/n!$ because photons are Bosons and have eliminated the polarization vectors by use of the equations for polarizations sums derived earlier in the chapter. However, in constructing the quantity A we must sum over all the possible external electron lines. Let there be M external electron lines with momenta $p_1 \ldots p_M$. The quantity A is then the sum

$$\prod_{i=1}^{n} \frac{1}{(2\pi)^{3/2}(2|\mathbf{q}_i|)^{1/2}} \sum_{j=1}^{M} \zeta_j \frac{p_j \cdot \varepsilon(q_i)}{(p_j \cdot q_i)}. \tag{4.132}$$

Also B is replaced by

$$D_n = \frac{1}{n!} \prod_{i=1}^{n} \frac{1}{(2\pi)^3 2|\mathbf{q}_i|} \sum_{j,k}^{M} \zeta_j \zeta_k \frac{(p_j \cdot p_k)}{(p_j \cdot q_i)(p_k \cdot q_i)}. \tag{4.133}$$

Next we have to integrate D_n over the solid angles Ω_i of the soft photons q_i and over the range of energies for which we regard the photon momentum q_i to be soft. The integration over the solid angles is accomplished by, (remember these are real photons so $q_i^2 = 0$), evaluating the expression

$$E_n = \frac{1}{n!}\prod_{i=1}^{n}\int d^4q_i \frac{\delta(q_i^2)}{(2\pi)^3(2|\mathbf{q}_i|)}\sum_{j,k}^{M}\frac{-\zeta_j\zeta_k(p_j\cdot p_k)}{(p_j\cdot q_i)(p_k\cdot q_i)}. \qquad (4.134)$$

However E_n is still not the final expression for we have not incorporated any function in the integrand which requires the photons to be soft. What we do is to restrict the maximum energy of all the soft photons to be $\leq \Delta E$ where ΔE is a small energy. In fact ΔE is an experimental limitation of the experimental apparatus: it is the smallest energy photon that we can detect. Since the aim is to extract from the integral an infrared divergent quantity it is also sensible not to allow the photon energies to be 0 in the integrand. In fact we restrict the minimum energy of the photons to be $\geq \lambda$. The infrared divergence is then manifest as λ approaches zero. The upper energy limit ΔE on the energy of the soft photon cloud, $\sum_{i=1}^{n} q_{i0}$, is accomplished by using a step function $T(q_{i0}, \Delta E)$ where T has the standard integral representation

$$T(q_{i0}, \Delta E) = \frac{1}{\pi}\int dx \frac{\sin(\Delta E x)}{x}\exp\left[i\sum_{i=1}^{n} q_{i0}x\right]. \qquad (4.135)$$

The final expression for the contribution to the cross-section is

$$I = \sum_{0}^{\infty}\frac{1}{n!}\frac{1}{\pi}\int dx \frac{\sin(\Delta E x)}{x}\exp\left[i\sum_{i=1}^{n} q_{i0}x\right]\cdot$$

$$\prod_{i=1}^{n}\int_{\lambda}^{\Delta E} d^4q_i \frac{\delta(q_i^2)}{(2\pi)^3 2|\mathbf{q}_i|}\sum_{j,k}^{M}\frac{-\zeta_j\zeta_k(p_j\cdot p_k)}{(p_j\cdot q_i)(p_k\cdot q_i)}. \qquad (4.136)$$

We have summed n from 0 to ∞ as is necessary since, as we have shown above, there is no upper limit on the photon number. It is easily seen that the sum which defines I exponentiates so that I may be written

$$I = \pi^{-1}\int dx \frac{\sin(\Delta E x)}{x}\exp\left[P\int_{\lambda}^{\Delta E}\frac{dq_0}{q_0}e^{iq_0 x}\right] \qquad (4.137)$$

where P is given by the equation

$$P \ln\left(\frac{\Delta E}{\lambda}\right) = \frac{1}{2(2\pi)^3} \int_\lambda^{\Delta E} d^4q\, \delta(q^2) \sum_{j,k}^M \zeta_j\zeta_k \frac{-(p_j \cdot p_k)}{(p_j \cdot q)(p_k \cdot q)}. \quad (4.138)$$

The integral in 4.143 is completely straightforward and we can immediately deduce that

$$P = -\frac{1}{8\pi^2} \sum_{j,k}^M \zeta_j\zeta_k [1 - m^4/(p_j \cdot p_k)^2]^{-1/2} \ln\left\{\frac{1 + [1 + m^4/(p_j \cdot p_k)^2]^{1/2}}{1 - [1 - m^4/(p_j \cdot p_k)^2]^{1/2}}\right\}. \quad (4.139)$$

The quantity $[1 - m^4/(p_j \cdot p_k)^2]^{1/2}$ has a physical interpretation: it is the relative velocity of the electrons p_j and p_k. We now wish to estimate the behaviour of I as λ approaches zero. For this we simply need to observe that

$$\lim_{\lambda \to 0} \int_\lambda^{\Delta E} \frac{dq_0}{q_0} e^{iq_0 x} \to \ln(\Delta E/\lambda) + \int_0^{\Delta E} \frac{dq_0}{q_0}(e^{iq_0 x} - 1) + \cdots. \quad (4.140)$$

Hence we have

$$\lim_{\lambda \to 0} I \to \left(\frac{\Delta E}{\lambda}\right)^P \frac{1}{\pi} \int dx \frac{\sin(\Delta E x)}{x} \exp\left[P \int_0^{\Delta E} \frac{dq_0}{q_0}(e^{iq_0 x} - 1)\right]. \quad (4.141)$$

The constant P is positive so that I goes to infinity as λ goes to zero. Now, according to the reasoning at the beginning of this analysis, I is a factor that multiplies the non-infrared divergent part of the cross-section σ when radiation from external soft photons is taken into account. We summarize this in the equation

$$\sigma_{\text{ext}} = \left\{\frac{\Delta E}{\lambda}\right\}^P B\sigma \quad (4.142)$$

where

$$B = \frac{1}{\pi} \int dx \frac{\sin(\Delta E x)}{x} \exp\left[P \int_0^{\Delta E} \frac{dq_0}{q_0}(e^{iq_0 x} - 1)\right]. \quad (4.143)$$

Equation 4.142 is the cross-section corrected for radiated soft photons of total energy $\leq \Delta E$. We now finish the infrared calculation by correcting the cross-section for insertion of internal soft photons. In Fig. 6(b) we show an internal soft photon connecting two external lines. In fact we shall not need to consider soft photons that connect lines other than external lines. This is because internal electron lines have momenta that are not on or near their mass shells and so do not give rise to infrared divergences. Our calculation proceeds along similar lines to that for external soft photons. The insertion of a soft photon between lines of momentum p_1 and p_2 gives rise to the factor

$$\frac{-i}{(q^2+i\varepsilon)} \zeta_1 \zeta_2 \frac{(p_1 \cdot p_2)}{(p_1 \cdot q)(p_2 \cdot q)} \tag{4.144}$$

where q is the momentum of the photon. We say that the momentum of the internal photon is soft if $|q| \geq \lambda$, and $|q|$ is less than some upper limit, Λ.

If n soft photons are inserted into the core diagrams and all permutations are summed over, the result is the factor

$$E_n = \frac{1}{n!} \left[\frac{1}{2} \int_\lambda^\Lambda \frac{d^4q}{(2\pi)^4} \frac{-i}{(q^2+i\varepsilon)} \sum_{j,k}^M \frac{\zeta_j \zeta_k (p_j \cdot p_k)}{(p_j \cdot q)(-p_k \cdot q)} \right]^n. \tag{4.145}$$

The factor $\frac{1}{2}$ is needed for the two directions of flow of the charge along the electron lines. The symbol $\int_\lambda^\Lambda (d^4q/(2\pi)^4)$ is the loop integral with the upper and lower limits included to keep the photon soft. The sum of E_n from 0 to ∞ produces the factor

$$J = \exp\left[\frac{1}{2} \int_\lambda^\Lambda d^4q \, Q(q) \right] \tag{4.146}$$

where

$$Q(q) = \frac{-i}{(2\pi)^4 (q^2+i\varepsilon)} \sum_{j,k}^M \frac{\zeta_j \zeta_k (p_j \cdot p_k)}{(p_j \cdot q)(-p_k \cdot q)}. \tag{4.147}$$

The contribution of J to the cross-section is given by

$$\bar{I} = |J^2| = \exp\left[\mathrm{Re} \int_\lambda^\Lambda d^4q \, Q(q) \right] \tag{4.148}$$

(Re means real part). The real part is obtained by noting that

$$\mathrm{Re} \, \frac{i}{(q^2+i\varepsilon)} = \pi \delta(q^2). \tag{4.149}$$

Therefore we find that

$$\bar{I}' = \exp\left[\frac{1}{2(2\pi)^3} \int_\lambda^\Lambda d^4q \, \delta(q^2) \sum_{j,k}^M \frac{\zeta_j \zeta_k (p_j \cdot p_k)}{(p_j \cdot q)(p_k \cdot q)} \right]. \tag{4.150}$$

Fortunately we have already evaluated the integral in 4.150 so that using 4.138 we have

$$\bar{I} = \exp\left[-P \ln\left(\frac{\Lambda}{\lambda}\right)\right]$$

$$= \left(\frac{\lambda}{\Lambda}\right)^P. \tag{4.151}$$

(note that $I = 0$ when $\lambda = 0$ since P is positive). Finally the cross-section fully corrected for the insertion of all kinds of soft photons is given by

$$\sigma_{\text{ext,int}} = \left(\frac{\lambda}{\Lambda}\right)^P \left(\frac{\Delta E}{\lambda}\right)^P B\sigma$$

$$= \left(\frac{\Delta E}{\Lambda}\right)^P B\sigma. \tag{4.152}$$

We see that the promised cancellation has occurred: $\sigma_{\text{ext,int}}$ is independent of λ and is thus infrared finite. The two parameters ΔE and Λ remain in 4.152. We have already explained that ΔE is the resolution of the experimental device for detecting photons. Λ is an ultraviolet cut-off parameter; it may be absorbed into the infrared finite cross-section σ to be taken care of when renormalization removes the ultraviolet infinities. For example, if a convenient normalization mass M is chosen we may write

$$\sigma_{\text{ext,int}} = \left(\frac{\Delta E}{M}\right)^P \left(\frac{M}{\Lambda}\right)^P B\sigma$$

$$= \left(\frac{\Delta E}{M}\right)^P \bar{\sigma} \tag{4.153}$$

with the obvious definition for $\bar{\sigma}$. $\bar{\sigma}$ is then the quantity in which only ultraviolet infinities remain.

We finish this chapter by reminding the reader that it is the cross-sections rather than the S-matrix elements which are infrared finite. This is in contrast with the situation for the ultraviolet infinities where we find that an S-matrix element $S_{\alpha\beta}$ is finite after renormalization.

References

1. R. P. Feynman, *Acta Physika Polon.*, **24**, 697 (1963); L. D. Fadeev and V. N. Popov, *Phys. Lett.*, **25B**, 29 (1967); A. Salam and J. Strathdee, *Nuovo Cimento*, **11A**, 397 (1972).

2. G. 't Hooft and M. Veltman, *Diagrammar CERN* (1973).
3. S. S. Schweber, 'An Introduction to Relativistic Quantum Field Theory', Harper and Row (1971).
4. R. J. Eden, P. V. Landshoff, D. I. Olive and J. C. Polkinghorne, 'The Analytic S-Matrix', Cambridge University Press (1966).
5. *cf. papers cited in* Nicolas Papenicolaou, *Phys. Reports*, **24C**, No. 4 (1976).
6. T. W. Kibble, *J. Math. Phys.*, **9**, 315 (1968); *Phys. Rev.*, **173**, 1527 (1968); **174**, 1882 (1968); **175**, 1624 (1968).
7. S. Weinberg, *Phys. Rev.*, **140B**, 516 (1965).

FIVE

Asymptotic Behaviour and Renormalization Group Methods

Asymptotic Behaviour of Scattering Amplitudes

In this chapter we wish to examine the asymptotic behaviour of Green's functions in perturbation theory. This subject has always been one of great interest to elementary-particle theorists. The reason is that for large values of the energies of the particles involved in a scattering process certain simplifying features are expected to be noticed. These features have been widely discussed and researched in the past twenty years or so but the investigations can by no means be regarded as complete. Since the advent of the isospin group SU(2) Lie groups have become of immense importance in elementary-particle physics. The larger symmetry groups SU(3), SU(6), SU(3)⊗SU(3), etc. have become commonplace in the literature. The so-called current-algebra at infinite momentum[1] was an important step in using symmetry groups to understand and formulate the asymptotic behaviour of scattering amplitudes at high energies. Another particularly important theoretical and experimental development tied closely to asymptotic behaviour is Regge theory and duality theory.[2] At very large energies and momentum transfers the four-momenta-squared of particles becomes very much larger than their rest mass. This leads to the suggestion that at such energies the physics is governed by a kind of zero-mass theory. This zero-mass theory is a limit of the original theory in which all the particles are given zero mass. The scattering amplitudes of this sort of idealized theory might be expected to have a chance of being invariant under a larger kinematic group than the Poincaré group: the conformal group.[3] This leads to many interesting and tantalizing results.[3] There is also, of course, the work of dispersion theory[4] which is of vital importance in pinning down the details of scattering amplitudes, phase shifts and particle spectra. Much of this work relies on experimental measurements made at high energies and theoretical knowledge of asymptotic behaviours. More recently there has been the experimental discovery of approximate scaling in deep inelastic scattering.[5] This has given rise to several theoretical lines of research. We give the names and references for some of these. There is the fundamental short distance expansion proposed by Wilson,[6] and the more ambitious light cone expansion of Brandt and Preparata,[7] and also Frishman.[7] Then there is a generalization of the equal time algebra of currents[1] to the light-cone algebra of Fritzsch and Gell-Mann.[8] Finally there are the renormalization group methods of Callan, Symanzik, 't Hooft and Weinberg;[9] and the important property known as asymptotic freedom.[10] In this last chapter we cannot deal with all these topics, so our attention will for the most part be focused on the renormalization group methods.

Asymptotic behaviours of individual Feynman diagrams or sums of Feynman diagrams can be calculated more or less directly by a number of methods. These methods have had great success particularly in the Regge limit. The techniques used involve exploiting the analyticity structure of the scattering amplitudes which obtain in perturbation theory, knowledge of the Landau singularities, and the use of Mellin transforms. The reader is referred to The Analytic S-matrix cited in reference 4 for an excellent account of these methods. The renormalization group methods are a more mechanized and less laborious way of calculating asymptotic behaviours. At the moment though they are a little limited in the kinematical regimes to which they can be usefully applied. However, one hopes that some of these limitations may be overcome.

Renormalization Group Methods

The history of the renormalization group goes back a long way.[11] A

ASYMPTOTIC BEHAVIOUR

minor source of worry to the reader may be that throughout this chapter no group will be introduced to justify the use of the term renormalization group. The fact is that the nomenclature is an unfortunate one. There is a one-parameter group of renormalizations but this is of no particular value in actual computations. A further source of complication is that there are really three types of renormalization group equation: the Gell-Man–Low equation, the Callan–Symanzik equations and the 't Hooft–Weinberg equations. The Gell-Mann–Low equation is confined to QED while the Callan–Symanzik equations and the 't Hooft–Weinberg equations are not. Further the basic results obtainable with the Gell-Man–Low equations are contained in the results obtainable with the Callan–Symanzik and 't Hooft–Weinberg equations. However, the Gell-Mann–Low equation provides us with a convenient place for beginning our discussion of renormalization group equations. We shall not derive the Gell-Mann–Low equation since we shall later on be deriving the other two renormalization group equations.

The Gell-Mann–Low Equation

Let $d(p^2)$ be the function defined in QED by

$$e^2 D_F^{\mu\nu}(p) = -\frac{g^{\mu\nu}}{p^2} d(p^2, e^2) \tag{5.1}$$

where $D_F^{\mu\nu}(p)$ is the renormalized photon propagator in the Feynman gauge and e is the renormalized charge which has the value given by $e^2/4\pi = (137)^{-1}$. Now by exploiting the arbitrariness discussed in Chapters 2 and 3 of the counter-term subtraction constants of QED Gell-Mann and Low obtained[11] an implicit equation for $d(p^2)$ when $p^2 \gg 0$. The equation took the form

$$\ln(p^2/M^2) = \int_{(e')^2}^{d_A(p^2, e'^2)} \frac{dz}{\psi(z)}. \tag{5.2}$$

In 5.2 $M \gg m$, m is the electron mass, d_A is the asymptotic form of d for large p^2 and e' is a new renormalized coupling constant differing by a finite renormalization from e. That is to say

$$e' = Z\left(\frac{M}{m}\right) e \tag{5.3}$$

where Z is finite. Lastly $\psi(z)$ is a function defined by an infinite power series which may be calculated by evaluating Feynman graphs in QED. The bold reasoning of Gell-Mann and Low concerned the asymptotic form d_A of d. In 5.2 as $p^2 \to \infty$ the LHS becomes infinite. There are two ways in which this might happen. The first is that the integrand ψ^{-1} remains finite throughout the range of integration but that the range of integration $(\epsilon')^2$, $d_A(p^2, (e')^2)$ stretches to infinity as $p^2 \to \infty$ in such a way as to make the integral diverge. Clearly this means that $d_A \to \infty$ as $p^2 \to \infty$, an entirely possible result. The second possibility is that the integrand does not remain infinite and that instead there is a value of $Z : e_{\text{GML}}$, for which

$$\psi(e_{\text{GML}}) = 0. \tag{5.4}$$

This infinity of ψ^{-1} is then meant to cause the infinity in the LHS. The asymptotic form of $d(p^2, e^2)$ for large p^2 is then given according to the Gell-Mann–Low equation 5.2 by $d_A(p^2, e_{\text{GML}}^2)$ and need not be infinite. This exciting possibility rests on discovering whether there is a solution e_{GML} to 5.4. But $\psi(z)$ is an infinite power series of which only the first few terms are known thus little progress has been made in this direction. Nevertheless we can take a lesson from the Gell-Mann–Low equation. It is that the asymptotic behaviour of a Green's function in perturbation theory (in this case the photon propagator of QED) can be determined by knowledge of a single function of the coupling constant. This lesson will also be learned from a study of the Callan–Symanzik and 't Hooft–Weinberg equations.

The Callan–Symanzik Equations

We shall derive the Callan–Symanzik equation for $\lambda \phi^4/4!$ theory. The derivation for other renormalizable theories follows exactly similar lines.† Let L be the Lagrangian for the renormalized theory introduced at the end of Chapter 2. Then consider a family of Lagrangians L' where

$$L' = L - s Z_3 \frac{\delta m^2 : \phi^2 :}{2} \tag{5.5}$$

† For theories with a gauge invariance some extra remarks are needed. We shall supply these later in the chapter.

and s is a real number. Evidently L' is the same Lagrangian as L but with a mass counter-term differing by a finite amount proportional to s. However, both Lagrangians L' and L are equally good candidates for describing a physical world in which ϕ interacts as $\lambda\phi^4/4!$. But if the mass counter-term is changed so will the resultant renormalized mass. So in its turn will be the renormalized coupling constant. This is because mass counter-terms must be used in the Feynman graphs which contribute to the four-point function which is used to define the renormalized coupling constant. Let us denote an arbitrary one particle irreducible Green's function calculated using the Lagrangian L by $\Gamma(p_1 \ldots p_n, m^2, \lambda)$. Let us denote the corresponding Green's function calculated using L' by $\Gamma'(p_1 \ldots p_n, m^2(s), \lambda(s))$ where $m(s)$ and $\lambda(s)$ are the corresponding renormalized parameters. Because the differences in the Lagrangians L and L' is only a finite change of counter-term the two sets of Green's functions are simply related. The relation is given by

$$[Z(s)]^{n/2} \Gamma(p_1 \ldots p_n, m^2, \lambda) = \Gamma'(p_1 \ldots p_n, m^2(s), \lambda(s)) \quad (5.6)$$

where $Z(s)$ is a finite function. The reason that the function $Z(s)$ is finite is that Γ' and Γ are both renormalized Green's functions and are finite by construction. Hence in a relation between them such as 5.6 the quantity $Z(s)$ must be finite also. The equation 5.6 may be derived fairly simply. Let $\Gamma_0(p_1 \ldots p_n, m_0^2, \lambda_0)$ be the bare one-particle irreducible Green's function. As usual the suffix 0 denotes the bare quantities. We know that the renormalized Green's functions are multiplicatively related to the bare Green's functions. For example in this particular set of circumstances we necessarily have the equations

$$\Gamma(p_1 \ldots p_n, m^2, \lambda) = [Z_3]^{n/2} \Gamma_0(p_1 \ldots p_n, m_0^2, \lambda_0) \quad (5.7)$$

and

$$\Gamma'(p_1 \ldots p_n, m^2(s), \lambda(s)) = [Z_3(s)]^{n/2} \Gamma_0(p_1 \ldots p_n, m_0^2, \lambda_0). \quad (5.8)$$

But this means that we may write

$$\Gamma'(p_1 \ldots p_n, m^2(s), \lambda(s)) = \left[\frac{Z_3(s)}{Z_3}\right]^{n/2} Z_3^{n/2} \Gamma_0(p_1 \ldots p_n, m_0^2, \lambda_0)$$

$$= \left[\frac{Z_3(s)}{Z_3}\right]^{n/2} \Gamma(p_1 \ldots p_n, m^2, \lambda). \quad (5.9)$$

Thus we have found that $Z(s)$ is given by the equation

$$Z(s) = \frac{Z_3(s)}{Z_3}. \tag{5.10}$$

The two sets of renormalized Green's functions Γ and Γ' are said to differ by a finite renormalization. We point out that even though $Z_3(s)$ and $Z(s)$ are separately infinite their ratio is finite by the argument we presented above. The reason that we may construct two different renormalized theories differing only in their finite parts is because of the indeterminacy of the finite part discussed in Chapter 2. We refer to the discussions following eqs 2.18 and 2.27. Returning to 5.6 we differentiate with respect to s and then set $s=0$. The equation that results is

$$\left[\alpha(\lambda)m^2\frac{\partial}{\partial m^2}+\beta(\lambda)\frac{\partial}{\partial \lambda}-n\gamma(\lambda)\right]\Gamma = \Delta\Gamma(p_1\ldots p_n, m^2, \lambda). \tag{5.11}$$

This equation is obtained merely by using the partial differentiation chain rule. The notations for the quantities introduced in 5.11 are easily seen to be:

$$\alpha(\lambda) = \frac{1}{m^2(s)}\frac{dm^2(s)}{ds}, \qquad s=0$$

$$\beta(\lambda) = \frac{d\lambda(s)}{ds}, \qquad s=0$$

$$\gamma(\lambda) = +\frac{1}{2Z(s)}\frac{dZ(s)}{ds}, \qquad s=0 \tag{5.12}$$

$$\Delta\Gamma = -\frac{d\Gamma'}{ds}, \qquad s=0.$$

The coefficient functions $\alpha(\lambda)$, $\beta(\lambda)$ and $\gamma(\lambda)$ are all power series in the coupling constant λ. We shall deal with these functions in a moment. The quantity $\Delta\Gamma$ has a simple relation to the Green's function Γ from which it was derived; $\Delta\Gamma$ consists of all those Feynman graphs which contribute to the Green's function Γ with the modification that all these graphs must have the mass counter-term $Z_3(\delta m^2 : \phi^2 :/2)$ inserted in all possible ways. It is straightforward to see why this is so. We saw in Chapter 1 that exp $[iS]$ where S is the action is the basic quantity for determining the Green's functions. Clearly the modification of L from L to L^s changes the action by the amount $\int dx\, s(:\delta m^2 \phi^2:/2)$. Hence the basic integrand in the functional integral changes from

$$\exp\left[i \int dx\, L\right] = I \tag{5.13}$$

to

$$\exp\left[i \int dx\, L + is: \int dx\, Z_3 \frac{\delta m^2 \phi^2}{2}:\right] = I^s. \tag{5.14}$$

The ultimate effect of differentiating with respect to s and setting $s = 0$ is to replace the integrand in the functional integral by

$$\frac{dI}{ds}, (s = 0) = iZ_3 \int dx\, :\frac{\delta m^2 \phi^2}{2}: \exp\left[i \int dx\, L\right]. \tag{5.15}$$

This means that the Green's functions obtained by taking moments with respect to ϕ of dI^s/ds, $s = 0$, rather than I, will be the quantities

$$\Delta G(x_1 \ldots x_n) = i \int dx\, \langle 0|T(Z_3\, \delta m^2 :\phi^2(x): \phi(x_1) \ldots \phi(x_n))|0\rangle \tag{5.16}$$

rather than

$$G(x_1 \ldots x_n) = \int dx\, \langle 0|T(\phi(x_1) \ldots \phi(x_n))|0\rangle. \tag{5.17}$$

The basic point to realize is that $\Delta\Gamma$ is constructed from Γ by inserting a multiple of the operator $:\phi^2(x):$.

Now we come to the coefficient functions $\alpha(\lambda)$, $\beta(\lambda)$ and $\gamma(\lambda)$. Firstly $\alpha(\lambda)$ may be eliminated by dividing both sides by $\alpha(\lambda)$. More precisely we are free to normalize the inserted mass term of $\Delta\Gamma$ so that $\alpha(\lambda) = 1$. The freedom to normalize $\Delta\Gamma$ is precisely the same freedom that is present when normalizing the renormalized Green's functions. That is to say the unrenormalized versions of these Green's functions Γ and $\Delta\Gamma$ are in general infinite and the finite parts are indeterminate until normalization conditions are imposed. The normalization conditions for $\Delta\Gamma$ and Γ that we require are

$$\Delta\Gamma(p, -p, m^2, \lambda) = -im^2; \qquad p^2 = m^2 \qquad (5.18)$$

$$\Gamma(p, -p, m^2, \lambda) = 0; \qquad p^2 = m^2; \qquad \frac{\partial \Gamma}{\partial p^2}(p, -p, m^2, \lambda) = i;$$

$$p^2 = m^2 \qquad (5.19)$$

$$\Gamma(p_1 \ldots p_4, m^2, \lambda) = -i\lambda, \qquad p_i \cdot p_j = \frac{m^2}{3}(4\delta_{ij} - 1), \qquad i,j = 1 \ldots 4.$$

$$(5.20)$$

The choice of normalization condition for $\Delta\Gamma$ has already been explained. The remaining conditions are simply a list of those that we used in Chapter 1. The first condition of 5.19 simply says that if the renormalized propagator Δ_F has a pole at the renormalized mass, then the inverse propagator has a zero there. The second condition of 5.19 simply fixes the residue of this pole, the renormalization constant having been factored out. The last condition (5.20) is the definition we used for the renormalized coupling constant. Only three normalization conditions are needed because there are three coefficient functions. The functions $\beta(\lambda)$ and $\gamma(\lambda)$ are infinite power series in λ

$$\begin{aligned} \beta(\lambda) &= b_0 \lambda^2 + b_1 \lambda^3 + \cdots, \\ \gamma(\lambda) &= d_0 \lambda^2 + d_1 \lambda^3 + \cdots. \end{aligned} \qquad (5.21)$$

We shall now show how to calculate the coefficients $b_0, b_1, \ldots$ and $d_0, d_1, \ldots$. They are calculated by combining the normalization conditions 5.18–5.20 with the partial differential equation (5.11). We can be sure at the outset that these coefficients will be finite. This is so because $\beta(\lambda)$ and $\gamma(\lambda)$ appear in an equation for the renormalized (and so finite) Green's functions Γ and $\Delta\Gamma$. Therefore they must themselves be finite. We shall give the procedure for calculating $\beta(\lambda)$ first. The idea is to utilize normalization condition 5.20. To be specific we set $n = 4$ in 5.11 and so obtain an equation for $\Gamma(p_1 \ldots p_4, m^2, \lambda)$

$$\left[m^2 \frac{\partial}{\partial m^2} + \beta(\lambda) \frac{\partial}{\partial \lambda} - 4\gamma(\lambda) \right] \Gamma(p_1, \ldots p_4, m^2, \lambda)$$

$$= \Delta\Gamma(p_1, \ldots p_4, m^2, \lambda). \qquad (5.22)$$

Then we evaluate both sides of 5.22 at the values of the momenta $p_1 \ldots p_4$ given in normalization condition 5.20, i.e. at the symmetry point. But at the symmetry point we have

$$\Gamma_{SP}(p_1, \ldots p_4, m^2, \lambda) = -i\lambda \qquad (5.23)$$

where SP stands for symmetry point. Thus 5.22 evaluated at the symmetry point gives the equation

$$m^2 \frac{\partial \Gamma_{SP}}{\partial m^2}(p_1, \ldots p_4, m^2, \lambda) - i\beta(\lambda) + 4i\lambda\gamma(\lambda) = \Delta\Gamma_{SP}(p_1, \ldots p_4, m^2, \lambda). \qquad (5.24)$$

Hence

$$\beta(\lambda) = -im^2 \frac{\partial \Gamma_{SP}}{\partial m^2} + 4\lambda\gamma(\lambda) + i\Delta\Gamma_{SP}. \qquad (5.25)$$

This is our expression for $\beta(\lambda)$. We see that $\beta(\lambda)$ may be expressed as a power series in λ. This is because Γ and $\Delta\Gamma$ are sums of Feynman diagrams and $\gamma(\lambda)$ will also turn out to be a power series in λ. However, to lowest order in λ the term $\lambda\gamma(\lambda)$ will not contribute since it is of order λ^3 while the first Feynman diagrams contributing to Γ and $\Delta\Gamma$ are of order λ^2. Thus to this lowest order we are only required to calculate the quantity

$$-i\left[m^2 \frac{\partial \Gamma_{SP}}{\partial m^2} - \Delta\Gamma_{SP}\right]. \qquad (5.26)$$

The lowest order graphs which contribute to 5.26 are shown in Fig. 1.

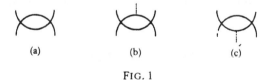

FIG. 1

There are two more sets of graphs, of course, corresponding to the two other ways of attaching the external legs: cf. eq. 2.22 of Chapter 2. The graph of order λ shown in Fig. 2 does not contribute. This is because it vanishes on application of $\partial/\partial m^2$; and it has no internal propagators to attach mass counter-terms to. The vertex with the dotted line shown in Fig. 1 requires explanation. It is the vertex that

FIG. 2

denotes the mass counter-term which is involved in the definition of $\Delta\Gamma$. Because of normalization condition 5.18 we can deduce that the bare vertex of Fig. 3 stands for the factor $-im^2$ in a general Feynman

FIG. 3

diagram. Note that the dotted line carries zero momentum. Thus apart from the factor $-im^2$ the effect of the mass insertion is to 'double up' the propagator to which it is attached. An internal line of momentum p has its denominator changed from $(p^2 - im^2)^{-1}$ to $(p^2 - m^2)^{-2}$. The application of the operation $m^2(\partial/\partial m^2)$ has a similar effect. An important reminder is now necessary. In the calculation of Fig. 1(a) a symmetry factor of $\frac{1}{2}$ is needed. In calculating Figs 1(b) and (c) the symmetry factor is $\frac{1}{2}$ because of the symmetry between the two internal lines in these graphs. In our discussion of renormalization we pointed out that Green's functions have a fixed dimension. In fact, $\Gamma(p_1 \ldots p_n, m^2, \lambda)$ has the dimension $(\text{mass})^{4-n}$. We now want to use this fact to facilitate our calculation of $\beta(\lambda)$. Let μ be a dimensionless parameter. Then for $\Gamma(p_1 \ldots p_4, m^2, \lambda)$ we have (remember here $n = 4$)

$$\Gamma(\mu p_1, \ldots \mu p_4, \mu^2 m^2, \lambda) = \Gamma(p_1, \ldots p_4, m^2, \lambda) \quad (5.27)$$

or

$$\Gamma(\mu p_1, \ldots \mu p_4, m^2, \lambda) = \Gamma\left(p_1, \ldots p_4, \frac{m^2}{\mu^2}, \lambda\right). \quad (5.28)$$

Differentiating with respect to μ^2 and setting $\mu = 1$ we have

$$\mu^2 \frac{\partial \Gamma}{\partial \mu^2}(\mu p_1, \ldots \mu p_4, m^2, \lambda) = -m^2 \frac{\partial \Gamma}{\partial m^2}\left(p_1, \ldots p_4, \frac{m^2}{\mu^2}, \lambda\right), \quad \mu = 1.$$

(5.29)

ASYMPTOTIC BEHAVIOUR

Thus we may replace 5.26 by

$$-i\left[-\mu^2\frac{\partial\Gamma_{\text{SP}}(\mu p_1,\ldots\mu p_4, m^2,\lambda)}{\partial\mu^2}-\Delta\Gamma_{\text{SP}}\right]. \quad (5.30)$$

Now if we refer to eq. 2.8 of Chapter 2 we find that (with $\mu = 1$, $s = \frac{4}{3}m^2$) that we may write 5.30 as

$$-i\left[\frac{\lambda^2}{2(2\pi)^4}(-\mu^2)\frac{\partial}{\partial\mu^2}\int dk\frac{d\alpha}{[k^2-m^2+\alpha(1-\alpha)\mu^2 s]^2}\right.$$
$$-\frac{(-i\lambda)^2}{(2\pi)^4}\frac{1}{2}\int\frac{dk(-im^2)i^3}{[k^2-m^2]^2[\{k-\mu p_1-\mu p_2\}^2-m^2]}$$
$$\left.-\frac{(-i\lambda)^2}{(2\pi)^4}\frac{1}{2}\int\frac{dk(-im^2)i^3}{[k^2-m^2][\{k-\mu p_1-\mu p_2\}^2-m^2]^2}\right]. \quad (5.31)$$

In the last two terms of 5.31 we have left explicit all the various multiplicative factors that occur in the Feynman rules to enable the reader to follow this first use of the mass insertion vertex of Fig. 3; the variable $s = (p_1+p_2)^2$. At the SP we see that $s = \frac{4}{3}m^2$. Straightforward algebra of the kind employed in Chapter 2 allows us to add together the terms in 5.31 and to obtain the expression

$$-\frac{i\lambda^2}{(2\pi)^4}\int dk\int_0^1 d\alpha\frac{[\alpha(1-\alpha)s-m^2]}{[k^2-m^2+\alpha(1-\alpha)s]^3}. \quad (5.32)$$

Note that this is a finite quantity as it should be since we argued that $\beta(\lambda)$ was finite. Performing the k integration gives the expression

$$-\frac{i\lambda^2}{(2\pi)^4}\frac{\Gamma(1)}{\Gamma(3)}i\pi^2\int_0^1 d\alpha\frac{[\alpha(1-\alpha)s-m^2]}{[\alpha(1-\alpha)s-m^2]}=\frac{\lambda^2}{2^5\pi^2}. \quad (5.33)$$

Finally we must include the two further sets of graphs of eq. 2.22. These graphs, of course, provide an identical contribution because of the symmetry between s, t and u at the symmetry point. Hence their inclusion simply has the effect of multiplying the RHS of 5.33 by 3. So we may finish our calculation of $\beta(\lambda)$ by writing

$$\beta(\lambda)=b_0\lambda^2+b_1\lambda^3+\cdots$$

where

$$b_0=\frac{3}{2^5\pi^2}. \quad (5.34)$$

It is clear that to calculate $\beta(\lambda)$ to higher orders in λ requires the inclusion of higher order Feynman graphs. In general then we cannot hope to know completely the power series which determines β. We must be content with the first few terms. We shall see later that the first term b_0 has a special significance in determining asymptotic behaviour.

We now turn to the calculation of $\gamma(\lambda)$. We begin by using the dimensional device used in 5.27. When $n = 2$ we have

$$\Gamma(\mu p, -\mu p, m^2, \lambda) = \mu^2 \Gamma\left(p, -p, \frac{m^2}{\mu^2}, \lambda\right). \tag{5.35}$$

Differentiating with respect to μ^2 and setting $\mu = 1$ we conclude that

$$p^2 \frac{\partial \Gamma}{\partial p^2}(p, -p, m^2, \lambda) = \Gamma(p, -p, m^2, \lambda) - m^2 \frac{\partial \Gamma}{\partial m^2}(p, -p, m^2, \lambda). \tag{5.36}$$

This means that we may rewrite the Callan–Symanzik equation for $n = 2$ as

$$\left[-p^2 \frac{\partial}{\partial p^2} + \beta \frac{\partial}{\partial \lambda} - 2\gamma + 1\right] \Gamma(p, -p, m^2, \lambda) = \Delta \Gamma(p, -p, m^2, \lambda). \tag{5.37}$$

The next thing to do is to make use of the normalization conditions 5.18, 5.19. It is not good enough to simply evaluate 5.37 at $p^2 = m^2$ for $\gamma(\lambda)$ disappears from the resultant expression which is

$$-im^2 = -im^2, \tag{5.38}$$

i.e. it is an identity. The correct expression is obtained by first differentiating 5.37 with respect to p^2 and then setting $p^2 = m^2$. The result of this is the equation

$$-\frac{\partial \Gamma}{\partial p^2} - p^2 \left(\frac{\partial^2}{\partial p^2}\right)^2 \Gamma + \beta \frac{\partial}{\partial \lambda} \frac{\partial \Gamma}{\partial p^2} - 2\gamma \frac{\partial \Gamma}{\partial p^2} + \frac{\partial \Gamma}{\partial p^2} = \frac{\partial \Delta \Gamma}{\partial p^2}, \quad p^2 = m^2. \tag{5.39}$$

Use of the normalization conditions 5.18, 5.19 reduces this to the equation

$$-m^2 \left(\frac{\partial^2}{\partial p^2}\right)^2 \Gamma \bigg|_{p^2 = m^2} = 2i\gamma(\lambda) + \frac{\partial \Delta \Gamma}{\partial p^2}, \quad p^2 = m^2$$

or

$$\gamma(\lambda) = \frac{i}{2}\frac{\partial \Delta \Gamma}{\partial p^2} + \frac{im^2}{2}\left(\frac{\partial^2}{\partial p^2}\right)^2 \Gamma, \quad p^2 = m^2. \quad (5.40)$$

This is our expression for $\gamma(\lambda)$. The calculation proceeds in an exactly similar fashion to that for $\beta(\lambda)$. All the integrals that will be required are given in Chapter 2. The result of doing the algebra is that

$$\gamma(\lambda) = d_0 \lambda^2 + d_1 \lambda^3 + \cdots$$

with

$$d_0 = \frac{1}{3 \cdot 2^{11} \pi^4}. \quad (5.41)$$

Note that the lowest term in $\gamma(\lambda)$ is of order λ^2. This allows us to justify our discarding the term $\lambda \gamma(\lambda)$ in calculating b_0 above. We have now derived the Callan–Symanzik equation and established some of the properties of its coefficient functions $\beta(\lambda)$ and $\gamma(\lambda)$, therefore we are ready to proceed to extract asymptotic behaviours of Green's functions. We begin by discussing the simplest sorts of asymptotic behaviour. The technique used is to take the Green's function $\Gamma(p_1, \ldots p_n, m^2, \lambda)$ and multiply all the momenta by μ and take μ large, i.e. consider

$$\lim_{\mu \to \infty} \Gamma(\mu p_1, \ldots \mu p_n, m^2, \lambda). \quad (5.42)$$

We also require that the $\lim \Gamma(\mu p_1, \ldots \mu p_4, m^2, \lambda)$ contains no exceptional momenta. Exceptional momenta are a little technical to explain. The reader is referred to Weinberg[12] where exceptional momenta are introduced to calculate asymptotic behaviours of Green's functions in perturbation theory. For technical reasons, partly due to problems of conditional convergence in the sign of the square of Minkowskian momenta, Weinberg[12] gave his results in Euclidean space. In Euclidean space the momenta $p_1 \ldots p_n$ of $\Gamma(p_1 \ldots p_n, m^2, \lambda)$ are said to be non-exceptional if no non-trivial partial sum $p_{i_1} + \cdots p_{i_r}$ vanishes. (A trivial partial sum which vanishes would be $p_1 + \cdots p_n$ which vanishes by energy momentum conservation.) This restriction to non-exceptional momenta provides the simplest examples of calculations of asymptotic behaviour. We shall refer later to the corresponding analogue in Minkowski space (i.e. Minkowskian exceptional momenta). The intuitive content underlying non-exceptional momenta can be clarified somewhat. If the momenta are non-exceptional then no channel variable $(p_{i_1} + \cdots p_{i_j})^2$ can carry finite momentum, or equivalently, zero momentum. Also all the momenta-squared $(p_1^2, p_2^2 \ldots p_n^2)$ go to infinity as μ goes to infinity. This means that the value of the denominators of the Feynman propagators have become as large as possible due to the routeing of infinite momenta through them in all possible ways. Therefore such an asymptotic behaviour should correspond to the sharpest asymptotic decrease of a perturbation-theory Green's function. This asymptotic behaviour should be one of the simplest to calculate. On the other hand suppose we consider a forward amplitude with zero momentum-transfer in one channel. This would mean exceptional momenta since we would have $p + q = 0$ for some momenta, p, q. A physical example of exceptional momenta is provided by measurement of the total cross-section for two particles a and b. Since this involves, via unitarity,[4] the imaginary part of a forward amplitude; further, the particles involved are on their mass shells and have $p^2 = m^2$ not $p^2 = \infty$. There are also commonly measured what are called n-particle inclusive processes, where one or more particles are detected in the final state.[28] We depict these in 5.43.

total cross-section: $a + b \to X$, X undetected

$a + b \to C + X$, c detected, X undetected

n-particle inclusive:

$a + b \to c + d + X$, c, d detected, X undetected.

(5.43)

These n-particle inclusive processes involve exceptional momenta because they deal with on-shell particles. But they also involve forward channels just as the total cross-section does. This is because by extending the use of unitarity used in deriving the optical theorem, and using analyticity properties of perturbation theory,[4] it can be proved, to a certain standard of rigour,[28] that the n-particle inclusive cross-sections are given by certain discontinuities in the complex momentum plane of the amplitudes shown in Fig. 4.

In the diagrams of Fig. 4 the momenta on the left and right of each diagram enter symmetrically. This means that several channels

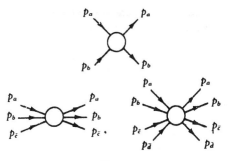

FIG. 4

carry zero momentum and therefore that we have exceptional momenta. We see then that exceptional momenta are the rule rather than the exception. Having drawn attention to this important point we return to the consideration of non-exceptional momenta. We write the Callan–Symanzik equation in the shortened form:

$$D\Gamma = \Delta\Gamma \qquad (5.44)$$

where D stands for the appropriate differential operator. Now for non-exceptional momenta $\Delta\Gamma(\mu p_1, \ldots, \mu p_n, m^2, \lambda)$ decreases more rapidly than $\Gamma(\mu p_1, \ldots \mu p_n, m^2, \lambda)$ as μ becomes large. To prove this statement one needs the details of the work in reference 12. However, we showed that the graphs of $\Delta\Gamma(p_1, \ldots p_n, m^2, \lambda)$ have always got a Feynman propagator which is doubled up. This means that the Feynman integral for these graphs will have dimension two less than the corresponding graph for $\Gamma(p_1, \ldots p_n, m^2, \lambda)$. When the momenta are non-exceptional there is always a large momentum routed through any doubled up propagator. This leads us to reason that, for non-exceptional momenta, the asymptotic behaviour of $\Delta\Gamma(\mu p_1, \ldots \mu p_n, m^2, \lambda)$ should be suppressed by a factor μ^2 compared with that of $\Gamma(\mu p_1, \ldots \mu p_n, m^2, \lambda)$. We can therefore neglect $\Delta\Gamma$ in 5.44 if we only wish to obtain the asymptotic form of $\Gamma(\mu p_1, \ldots \mu p_n, m^2, \lambda)$ for non-exceptional momenta. Let us call this asymptotic form Γ_{as}. Then we wish to solve

$$D\Gamma_{as} = 0, \qquad (5.45)$$

i.e.

$$\left[m^2 \frac{\partial}{\partial m^2} + \beta \frac{\partial}{\partial \lambda} - n\gamma\right]\Gamma = 0. \qquad (5.46)$$

Equation 5.46 is a linear homogeneous partial differential equation. It is therefore easy to solve. We change variables from m^2, λ to x, y where

$$x = \ln m^2, \qquad y(\lambda) = \int_{\lambda_0}^{\lambda} \frac{dz}{\beta(z)}. \qquad (5.47)$$

These variables x and y are chosen so that 5.46 becomes

$$\left[\frac{\partial}{\partial x} + \frac{\partial}{\partial y} - n\gamma\right]\Gamma_{as} = 0. \qquad (5.48)$$

This equation would be simple to solve if γ were zero. But the linear term $-n\gamma\Gamma$ can easily be eliminated in the beginning by multiplying by an integrating factor. Armed with this knowledge we immediately see that the general solution of 5.46 is

$$\Gamma_{as}(p_1, \ldots p_n, m^2, \lambda) = \exp\left[\int_{\lambda_0}^{\lambda} \frac{n\gamma(z)}{\beta(z)} dz\right] f^{p_1 \ldots p_n}(x - y)$$

$$= \exp\left[\int_{\lambda_0}^{\lambda} \frac{n\gamma(z)}{\beta(z)} dz\right] f^{p_1 \ldots p_n}\left(\ln m^2 - \int_{\lambda_0}^{\lambda} \frac{dz}{\beta(z)}\right). \qquad (5.49)$$

where f is any function. Clearly because of the complete freedom in choosing the function f, 5.49 does not contain much information about the underlying quantum field theory. Equation 5.49 is given substance by using it in conjunction with the identity

$$\Gamma_{as}(\mu p_1, \ldots, \mu p_n, \mu^2 m^2, \lambda) = \mu^{4-n}\Gamma_{as}(p_1, \ldots p_n, m^2, \lambda)$$

i.e.

$$\Gamma_{as}(\mu p_1, \ldots \mu p_n, m^2, \lambda) = \mu^{4-n}\Gamma\left(p_1, \ldots p_n, \frac{m^2}{\mu^2}, \lambda\right). \qquad (5.50)$$

This identity is, as we have said before, simply the statement that the one-particle irreducible Green's functions have the dimension $(\text{mass})^{4-n}$. Now if we combine solution 5.49 with 5.50 we obtain the result

$$\Gamma_{as}(\mu p_1, \ldots \mu p_n, m^2, \lambda) = \mu^{4-n} \exp\left[\int_{\lambda_0}^{\lambda} \frac{n\gamma(z)}{\beta(z)} dz\right]$$
$$\cdot f^{p_1 \ldots p_n}\left(\ln \frac{m^2}{\mu^2} - \int_{\lambda_0}^{\lambda} \frac{dz}{\beta(z)}\right). \qquad (5.51)$$

Next if we remember the equations for the characteristics of a partial differential equation we realize that we may absorb the $\ln \mu^2$ in the argument of f. More precisely define the function $\lambda(\mu)$, which we call an effective coupling constant, by

$$\lambda(\mu) = \lambda, \qquad \mu = 1, \qquad \ln \mu^2 = \int_\lambda^{\lambda(\mu)} \frac{d(z)}{\beta(z)}. \tag{5.52}$$

But 5.52, 5.47 imply that

$$y(\lambda(\mu)) = \int_{\lambda_0}^{\lambda(\mu)} \frac{dz}{\beta(z)} = \int_{\lambda_0}^{\lambda} \frac{dz}{\beta} + \int_\lambda^{\lambda(\mu)} \frac{dz}{\beta}$$
$$= y(\lambda) + \ln \mu^2. \tag{5.53}$$

So

$$f^{p_1 \ldots p_n}\left(\ln \frac{m^2}{\mu^2} - \int_{\lambda_0}^{\lambda} \frac{dz}{\beta}\right) = f^{p_1 \ldots p_n}\left(\ln m^2 - \ln \mu^2 - \int_{\lambda_0}^{\lambda} \frac{dz}{\beta}\right)$$
$$= f^{p_1 \ldots p_n}\left(\ln m^2 - \int_{\lambda_0}^{\lambda(\mu)} \frac{dz}{\beta(z)}\right). \tag{5.54}$$

But 5.54 is equal to

$$\exp\left[-n \int_{\lambda_0}^{\lambda(\mu)} \frac{\gamma}{\beta} dz\right] \Gamma_{as}(p_1, \ldots p_n, m^2, \lambda(\mu)) \tag{5.55}$$

by 5.49. Therefore 5.51 may be written as

$$\Gamma_{as}(\mu p_1, \ldots \mu p_n, m^2, \lambda) = \mu^{4-n} \exp\left[-n \int_\lambda^{\lambda(\mu)} dz \frac{\gamma}{\beta}\right]$$
$$\cdot \Gamma_{as}(p_1, \ldots p_n, m^2, \lambda(\mu)). \tag{5.56}$$

The formula of 5.56 is of great use and justifies the introduction of the effective coupling constant $\lambda(\mu)$. Equation 5.56 gives us a simple method of finding the asymptotic form of $\Gamma_{as}(p_1 \ldots p_n, m^2, \lambda)$ by evaluating $\Gamma_{as}(\mu p_1 \ldots \mu p_n, m^2, \lambda)$ for large μ. We simply write Γ_{as} with *finite momenta* $p_1 \ldots p_n$, but replace the coupling constant λ by the effective coupling constant $\lambda(\mu)$, and multiply by the exponential integrating factor which depends only on γ and β. When one examines 5.56 it becomes clear that asymptotic form for large μ depends crucially on how $\lambda(\mu)$ behaves for large μ. To answer this question we derive another equation for $\lambda(\mu)$. We differentiate 5.52 with respect to μ^2 and obtain the ordinary differential equation

$$\mu^2 \frac{d\lambda(\mu)}{d\mu^2} = \beta(\lambda(\mu)), \qquad \lambda(\mu) = \lambda, \qquad \mu = 1. \quad (5.57)$$

Simple calculus now allows us to answer the question about the behaviour of $\lambda(\mu)$. When the RHS $\beta(\lambda(\mu))$ is positive $\lambda(\mu)$ will increase with increasing μ and vice versa. We do not know β exactly but we know its first term. If we truncate the power series for β at its second term as an approximation we derive

$$\mu^2 \frac{d\lambda(\mu)}{d\mu^2} = b_0 \lambda^2(\mu), \qquad \lambda(\mu) = \lambda, \qquad \mu = 1. \quad (5.58)$$

The solution to this equation is

$$\lambda(\mu) = \frac{\lambda}{(1 - b_0 \lambda \ln \mu^2)}. \quad (5.59)$$

If μ is made large, but not so large that $b_0 \lambda \ln \mu^2 > 1$, then

$$\lambda(\mu) = \lambda(1 + b_0 \lambda \ln \mu^2 + b_0^2 \lambda^2 \ln^2 \mu^2 + \cdots). \quad (5.60)$$

As we would expect, since $b_0 > 0$, if μ increases $\lambda(\mu)$ increases. If we let μ increase so that $1 = b_0 \lambda \ln \mu^2$, $\lambda(\mu)$ is infinite. Clearly this cannot be allowed, as in allowing μ to be near that critical value $\lambda(\mu) - 1$ is large and positive and we would completely distrust our truncation of the series for β. Hence if we keep μ small enough we expect 5.58, 5.59, 5.60 to have some validity. We shall be discussing the significance of the sign of β in more detail later in the chapter. We are now ready to use 5.56 to derive an asymptotic behaviour. We choose to evaluate $\Gamma(\mu p_1, \ldots \mu p_4, m^2, \lambda)$ for large μ. Equation 5.56 gives for $n = 4$

$$\Gamma_{as}(\mu p_1, \ldots \mu p_4, m^2, \lambda) = \exp\left[-4 \int_\lambda^{\lambda(\mu)} dz \, \frac{\gamma(z)}{\beta(z)}\right]$$
$$\cdot \Gamma_{as}(p_1, \ldots p_4, m^2, \lambda(\mu)). \quad (5.61)$$

We deal first with the exponential factor. We have

$$\exp\left[-4\int_\lambda^{\lambda(\mu)} dz\, \frac{\gamma(z)}{\beta(z)}\right] = \exp\left[-4\int_\lambda^{\lambda(\mu)} dz\, \frac{(d_0 z^2 + \cdots)}{(b_0 z^2 + \cdots)}\right]$$

$$= \exp\left[-4\int_\lambda^{\lambda(\mu)} dz\, \frac{d_0}{b_0} + \cdots\right]$$

$$= \exp\left[-\frac{4 d_0 \lambda^2 \ln \mu^2}{(1 - b_0 \lambda \ln \mu^2)} + \cdots\right]$$

$$= 1 + O(\lambda(\mu)). \tag{5.62}$$

For the Green's function $\Gamma_{as}(p_1, \ldots p_4, m^2, \lambda(\mu))$ we have the perturbation expansion

$$\Gamma_{as}(p_1, \ldots p_4, m^2, \lambda(\mu)) = -i\lambda(\mu) + O(\lambda^2(\mu)). \tag{5.63}$$

Combining 5.62, 5.63 we conclude that for large μ (and, of course, non-exceptional momenta),

$$\Gamma_{as}(\mu p_1, \ldots \mu p_4, m^2, \lambda) = -i\lambda(\mu) \qquad (\mu \gg 0)$$

$$= \frac{-i\lambda}{(1 - b_0 \lambda \ln \mu^2)}$$

$$= \frac{-i\lambda}{(1 - (3\lambda/2^5 \pi^2) \ln \mu^2)}. \tag{5.64}$$

Equation 5.64 may be expanded as a power series in $\lambda \ln \mu^2$ and it then gives the correct asymptotic form for Γ_{as} as may be checked by direct calculation with sums of Feynman graphs. This example should be sufficient to illustrate the extremely simple working of the formula 5.56 for extracting asymptotic behaviours at non-exceptional momenta.

We would now like to relax the restriction of non-exceptional momenta and return to discussing the full Callan–Symanzik equation

$$D\Gamma = \Delta\Gamma. \tag{5.65}$$

We may write 5.65 as

$$\left[-\mu^2 \frac{\partial}{\partial \mu^2} + \beta \frac{\partial}{\partial \lambda} - n\gamma\right] \Gamma\left(p_1, \ldots p_n, \frac{m^2}{\mu^2}, \lambda\right) = \Delta\Gamma\left(p_1, \ldots p_n, \frac{m^2}{\mu^2}, \lambda\right). \tag{5.66}$$

We can view 5.66 as an inhomogeneous partial differential equation. Its general solution is then the sum of the general solution to the homogeneous equation and a particular solution to the inhomogeneous equation. The general solution to the homogeneous equation we already have. It is $\Gamma_{as}(p_1, \ldots p_n, m^2/\mu^2, \lambda)$. To obtain the particular solution we must first use the integrating factor and then the particular solution can be constructed by change of variables and straightforward integration. It can be checked by direct differentiation that the solution to 5.66 is

$$\Gamma\left(p_1, \ldots p_n, \frac{m^2}{\mu^2}, \lambda\right) = \Gamma_{as}\left(p_1, \ldots p_n, \frac{m^2}{\mu^2}, \lambda\right)$$
$$+ \int_{\mu^2}^{\infty} \frac{d\mu'^2}{\mu'^2} \exp\left[n \int_{\lambda(\mu/\mu')}^{\lambda} dz \frac{\gamma(z)}{\beta(z)}\right] \Delta\Gamma\left(p_1, \ldots p_n, \frac{m^2}{\mu'^2}, \lambda\left(\frac{\mu}{\mu'}\right)\right). \tag{5.67}$$

The asymptotic behaviour of $\Gamma(\mu p_1, \ldots \mu p_n, m^2, \lambda)$ is related to the LHS of 5.67 by 5.50. Now for non-exceptional momenta the second term in 5.67 ought to tend to zero. Let us see how this works. The second term in 5.67 is of the form

$$\int_{\mu^2}^{\infty} d\mu'^2 f(\mu', \mu) = I(\mu) \tag{5.68}$$

for an appropriate function $f(\mu', \mu)$. Clearly then, if the integral $I(\mu)$ exists for a general μ, $\lim_{\mu \to \infty} I(\mu)$ should be zero since the integration range is shrinking to zero during the limiting process. The only way around this argument is for the integral $I(\mu)$ not to exist. This is exactly the determining criteria for exceptional and non-exceptional momenta. For non-exceptional momenta $I(\mu)$ always exists and is zero for infinite μ so that in that case 5.67 becomes $\lim_{\mu \to \infty} \Gamma = \Gamma_{as}$.

The mechanism which gives rise to the non-existence of the integral $I(\mu)$ is the potentiality for $f(\mu', \mu)$ to diverge as μ becomes infinite. In other words $I(\mu)$ may diverge because a divergence of the integrand $f(\mu', \mu)$ is stronger than the convergence provided by the shrinking of the integration range to zero. Let us examine then the way in which $f(\mu', \mu)$ can diverge for large μ. The μ dependence of the integrand occurs in the quantities m^2/μ'^2 and $\lambda(\mu/\mu')$. Consider first $\lambda(\mu/\mu')$. We have seen, in, for instance, the example 5.60, that in perturbation theory $\lambda(\mu/\mu')$ grows as powers of logarithms for large

μ. This growth is not enough, in any finite order of perturbation theory, to provide an overall divergence of $I(\mu)$. Now we come to the quantity m^2/μ'^2. In the integral clearly we have $(\mu')^2 \geq \mu^2$ so for large μ, $m^2/(\mu')^2$ goes to zero. Thus the divergence that we are looking for is an infrared enhancement or divergence coming from a mass-parameter vanishing. We wish to know when $\Delta\Gamma(p_1, \ldots p_n, m^2, \lambda)$ is enhanced or divergent as m^2 goes to zero. The first part of the answer to this question is that it depends on the values of the external momenta. (We have seen in Chapter 4, for example, that Green's functions are only infrared divergent when external charged lines are on their mass shell.) The second half of the answer is simply that the infrared divergences of $\Delta\Gamma_{m\to 0}(p_1 \ldots p_n, m^2, \lambda)$ occur at exceptional momenta. Previously we argued that the doubling up of propagators in $\Delta\Gamma$ made it a more convergent object than $\Delta\Gamma$ asymptotically. However, for exceptional momenta this improved convergence is completely overcome by the infrared divergence. It is simple to give a concrete example of exceptional momenta at work. We derived above the asymptotic form of $\Gamma(\mu p_1, \ldots \mu p_4, m^2, \lambda)$ shown in 5.64. This derivation neglected $\Delta\Gamma$ in the Callan–Symanzik equation and so is valid only at non-exceptional momenta. Let us examine the lowest order graphs contributing to $\Delta\Gamma(\mu p_1, \ldots \mu p_4, m^2, \lambda)$ and allow the momenta to be exceptional so as to uncover an infrared divergence. The lowest order graphs are shown in Fig. 5.

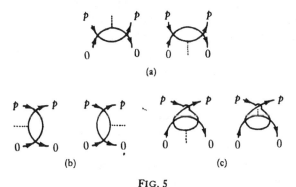

FIG. 5

The momenta are shown in the labelling of the external lines of Fig. 5; they are exceptional momenta according to the definition given above. We shall call the contributions from Figs 5(a), (b) and (c) $\Delta\Gamma(s)$, $\Delta\Gamma(t)$

and $\Delta\Gamma(u)$ respectively. The channel variables s, t, u have the values

$$s = (p+0)^2 = p^2$$
$$t = (p-p)^2 = 0,$$
$$u = (p+0)^2 = p^2.$$

Hence $\Delta\Gamma(s) = \Delta\Gamma(u) = \Delta\Gamma(p^2)$ and $\Delta\Gamma(t) = \Delta\Gamma(0)$ in this case. We have already written down the integrals for $\Delta\Gamma$ in 5.31. Reference to 5.31 yields the equations

$$\Delta\Gamma(s) = \frac{\lambda^2}{(2\pi)^4} \int dk \int_0^1 d\alpha \frac{m^2}{[k^2 - m^2 + \alpha(1-\alpha)s]^3}, \quad s = p^2;$$

$$\Delta\Gamma(t) = \frac{\lambda^2}{(2\pi)^4} \int dk \int_0^1 d\alpha \frac{m^2}{[k^2 - m^2]^3}, \quad t = 0;$$

$$\Delta\Gamma(u) = \frac{\lambda^2}{(2\pi)^4} \int dk \int_0^1 d\alpha \frac{m^2}{[k^2 - m^2 + \alpha(1-\alpha)u]^3}, \quad u = p^2.$$
(5.69)

We are interested in $\Delta\Gamma(\mu p, 0, \mu p, 0)$ for large μ. This simply means $s, u \to \infty$. Straightforward inspection of 5.32 shows that $\Delta\Gamma(s)$ and $\Delta\Gamma(u)$ go to zero as m^2/s, m^2/u, in this case. However $\Delta\Gamma(t)$ remains finite and constant since $t = 0$. In other words, for the set of momenta $p, 0, p, 0$, $\Delta\Gamma(\mu p, 0, \mu p, 0)$ is not negligible at large μ by virtue of the constancy of $\Delta\Gamma(t)$ at $t = 0$. Next we look at $\Delta\Gamma(p, 0, p, 0, m^2, \lambda)$ as m^2 goes to zero. If we perform the k integration in 5.69 we find that

$$\Delta\Gamma(s) = \Delta\Gamma(u) = \frac{i\lambda^2}{2^5 \pi^2} \int_0^1 d\alpha \frac{m^2}{[-m^2 + \alpha(1-\alpha)p^2]}. \quad (5.70)$$

Then as $m^2 \to 0$, since $p^2 \neq 0$, $\Delta\Gamma(s)$ and $\Delta\Gamma(u)$ go to zero. However $\Delta\Gamma(t)$ is simply $-i\lambda^2/2^5\pi^2$ and so does not go to zero for small m^2. Therefore $\Delta\Gamma(t)$ is enhanced relative to $\Delta\Gamma(s)$ and $\Delta\Gamma(u)$. This, of course, reflects the difference in asymptotic behaviour just discussed. Now if we insert $\Delta\Gamma(t)$ into 5.68 so as to obtain its contribution to $I(\mu)$ we obtain

$$\int_{\mu^2}^{\infty} \frac{d\mu'^2}{\mu'^2} \left(-\frac{i\lambda^2}{2^5 \pi^2}\right) \exp\left[n \int_{\lambda(\mu/\mu')}^{\lambda} dz \frac{\gamma(z)}{\beta(z)}\right]. \quad (5.71)$$

ASYMPTOTIC BEHAVIOUR 185

For large μ we can approximate this by

$$\int_{\mu^2}^{\infty} \frac{d\mu'^2}{\mu'^2}\left(-\frac{i\lambda^2}{2^5\pi^2}\right) = -\frac{i\lambda^2}{2^5\pi^2}[\ln \mu'^2]_{\mu^2}^{\infty} = \infty. \qquad (5.72)$$

In other words $\Delta\Gamma(t)$ provides a divergent contribution to $I(\mu)$ as claimed. This example is a typical illustration of how the infrared behaviour of $\Delta\Gamma(p_1, \ldots p_n, m^2, \lambda)$ at exceptional momenta shows that under these circumstances $\Delta\Gamma$ cannot be neglected in solving the Callan–Symanzik equations.

We argued above that from the physical viewpoint exceptional momenta are the interesting rather than the uninteresting cases to discuss. We would like then to be able to derive asymptotic behaviours in the exceptional cases as well. Our technique can be extended to do this by using the Wilson expansion[6] for operator products at short distances. We shall deal with the Wilson expansion in the next section of this chapter, but for now we shall just use the result without proof.

The problem is to find the asymptotic behaviour of $\Gamma(\mu p, 0, \mu p, 0)$ for large μ when Γ satisfies

$$D\Gamma(\mu p, 0, \mu p, 0) = \Delta\Gamma(\mu p, 0, \mu p, 0). \qquad (5.73)$$

As we have shown above by examining actual diagrams, $\Delta\Gamma(\mu p, 0, \mu p, 0)$ is not negligible for large μ. The Wilson expansion enables us to express the asymptotic behaviour of $\Delta\Gamma(\mu p, 0, \mu p, 0)$ in terms of $\Gamma(\mu p, 0, \mu p, 0)$. The result (that we prove in the next section) is that

$$\Delta\Gamma(\mu p, 0, \mu p, 0) \to \zeta(\lambda)\Gamma(\mu p, 0, \mu p, 0) + \cdots, \qquad \mu \to \infty. \qquad (5.74)$$

The dots in 5.74 stand for non-leading terms in the asymptotic expansion for large μ. The function $\zeta(\lambda)$ is a power series in λ and may be calculated to an arbitrary order in λ by use of the formula

$$im^2\zeta(\lambda) = \frac{m^4}{4}\int dx \, \langle 0|T(:\phi^2(x)::\phi^2(0):\hat{\phi}(0)\hat{\phi}(0)|0\rangle_{\bar{c}}$$

$$= e_0\lambda + e_1\lambda^2 + \cdots \qquad (5.75)$$

In 5.75 the symbol $\bar{c}$ has the same meaning as it did in Chapter 1: i.e. amputated connected Green's functions. Also $\hat{\phi}(p)$ is the Fourier transform of $\phi(x)$. With this formula for $\zeta(\lambda)$ one can calculate the coefficients $e_0, e_1, \ldots$ by evaluating Feynman diagrams just as one did for $\beta(\lambda)$ and $\gamma(\lambda)$. In this case the general Feynman diagram contributing to $\zeta(\lambda)$ will have two external 'ϕ legs', and two external ':ϕ^2: legs'. A general diagram is shown in Fig. 6(a). The general amplitude

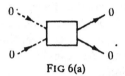

FIG 6(a)

represented in Fig. 6(a) has the expansion shown in Fig. 6(b). It is clear that e_0 comes from evaluating the diagram of order λ in Fig. 6(b). The

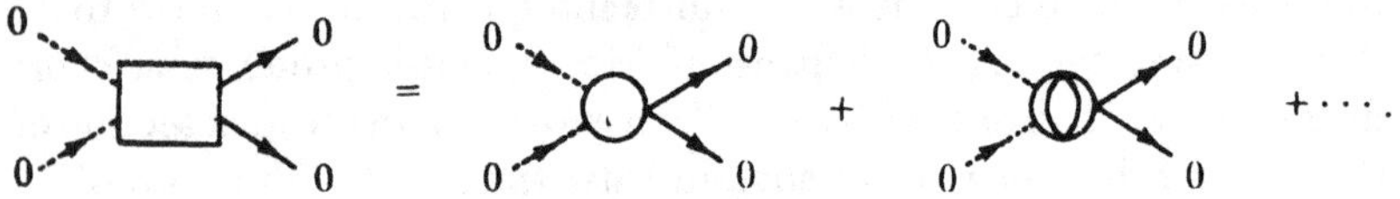

FIG. 6(b)

integral for this diagram is simply

$$\frac{\lambda m^4}{(2\pi)^4} \int \frac{dk}{[k^2 - m^2]^3} = e_0 \lambda(-im^2). \qquad (5.76)$$

This integral is trivial to evaluate and we obtain

$$\frac{i\lambda \pi^2 m^4 \Gamma(1)}{(2\pi)^4 (-m^2) \Gamma(3)} = e_0 \lambda(-im^2) \qquad (5.77)$$

or

$$e_0 = \frac{1}{2^5 \pi^2} = \tfrac{1}{3} b_0. \qquad (5.78)$$

We now return to the differential equation of eq. 5.73. If we write out the differential operator D in full and use 5.74 we have

$$\left[-\mu^2 \frac{\partial}{\partial \mu^2} + \beta \frac{\partial}{\partial \lambda} - 4\gamma \right] \Gamma^E_{as}(\mu p, 0, \mu p, 0)$$
$$= \zeta(\lambda) \Gamma^E_{as}(\mu p, 0, \mu p, 0) + \cdots. \qquad (5.79)$$

The notation Γ_{as}^E denotes an asymptotic form for exceptional momenta, but 5.79 may be written as the homogeneous equation

$$\left[-\mu^2\frac{\partial}{\partial\mu^2}+\beta\frac{\partial}{\partial\lambda}-4\gamma-\zeta\right]\Gamma_{as}^E(\mu p, 0, \mu p, 0) = 0. \qquad (5.80)$$

We have already solved this equation and obtained a convenient formula for writing down its behaviour for large μ. Comparison with 5.61 gives

$$\Gamma_{as}^E(\mu p, 0, \mu p, 0, m^2, \lambda)$$
$$= \exp\left[\int_\lambda^{\lambda(\mu)} dz \frac{(-4\gamma-\zeta)}{\beta}\right]\Gamma_{as}^E(p, 0, p, 0, m^2, \lambda(\mu)). \qquad (5.81)$$

The treatment of the exponential factor differs from the previous one of eq. 5.62. We have

$$\exp\left[\int_\lambda^{\lambda(\mu)} dz \frac{(-4\gamma-\zeta)}{\beta}\right] = \exp\left[\int_\lambda^{\lambda(\mu)} dz \frac{(-4d_0 z^2 - e_0 z + \cdots)}{(b_0 z^2 + \cdots)}\right]$$
$$= \exp\left[-\int_\lambda^{\lambda(\mu)} \frac{dz}{z}\frac{e_0}{b_0} + \cdots\right]$$
$$= \left[\frac{\lambda(\mu)}{\lambda}\right]^{-e_0/b_0} \qquad (5.82)$$

But $e_0 = \frac{1}{3}b_0$ thus 5.82 reduces to $[\lambda(\mu)/\lambda]^{(-1/3)}$. Returning to 5.81 we have

$$\Gamma_{as}^E(\mu p, 0, \mu p, 0, m^2, \lambda) = \left[\frac{\lambda(\mu)}{\lambda}\right]^{-1/3}[i\lambda(\mu) + \cdots]$$
$$= \left[1 - \frac{3\lambda}{2^5\pi^2}\ln\mu^2\right]^{1/3}\frac{(-i\lambda)}{[1-(3\lambda/2^5\pi^2)\ln\mu^2]}$$
$$= \frac{-i\lambda}{[1-(3\lambda/2^5\pi^2)\ln\mu^2]^{2/3}}. \qquad (5.83)$$

This is to be compared with the asymptotic form for non-exceptional momenta $\Gamma_{as}(\mu p_1, \mu p_2, \mu p_3, \mu p_4, m^2, \lambda)$ which is given in 5.64 as

$$-i\lambda(\mu) = \frac{-i\lambda}{[1-(3\lambda/2^5\pi^2)\ln\mu^2]}. \qquad (5.84)$$

We can check the validity of 5.83 by expanding it to second order in λ and comparing this with a direct calculation from Feynman diagrams. The configuration of momenta $\mu p, 0, \mu p, 0$ makes the order λ^2 contribution to the t-channel diagram finite for large μ^2 thus reducing the λ^2 contribution by $\frac{1}{3}$. This accounts for the reduction of the exponent from unity in eq. 5.84 to $\frac{2}{3}$ in eq. 5.83.

The Wilson Expansion at Short Distances

The Wilson expansion[6] investigates the behaviour of operator products $O(x)O(y)$ as the distance $(x-y)$ becomes small. Without knowing very much about the behaviour of operator products at short distances we can still say that in general we expect $O(x)O(y)$ to be singular as $x_\mu \to y_\mu$. For example in free field theory we know the product exactly. We have, (when $O(x) = \phi(x)$),

$$\phi(x)\phi(y) = \frac{1}{2\pi^2} \frac{1}{[(x-y)^2 + i\varepsilon(x_0 - y_0)]} \quad (5.85)$$

5.85 is singular when $x = y$ and also, of course, when $(x-y)^2 = 0$, i.e. on the light cone. However, we are not interested in the light cone singularity here. There is an extension of Wilson's short distance behaviours to light-cone behaviours.[7,8] We can also see that the general two-point function when ϕ is not a free field has a short distance singularity of some kind. We know that a general $\phi(x)$ satisfies the spectral representation*

$$\langle 0|\phi(x)\phi(y)|0\rangle = \int_0^\infty d\sigma^2 \, p(\sigma^2) i \, \Delta^+(x-y, \sigma^2) \quad (5.86)$$

as we derived the form of this representation for the photon field in Chapter 4. Clearly the behaviour of the LHS as $x \to y$ depends on the behaviour of $\Delta^+(x-y, \sigma^2)$ as $x \to y$. This behaviour is simple to calculate since $\Delta^+(x-y, \sigma^2)$ is a known function. The result is that

$$i\Delta^+(x-y, \sigma^2) \underset{x \to y}{\to} \frac{1}{2\pi^2[(x-y)^2 + i\varepsilon(x_0 - y_0)]} + \cdots. \quad (5.87)$$

This behaviour is independent of σ^2 and coincides with the free field behaviour. The rest of the analysis depends on the details of the

ASYMPTOTIC BEHAVIOUR 189

behaviour of the σ^2 integral over $\rho(\sigma^2)$. But we see that $\langle 0|\phi(x)\phi(y)|0\rangle$ is at least as singular as it is in the free field case. Wilson has proposed an asymptotic expansion for the product of operators at short distances. This asymptotic expansion has been established in perturbation theory by Zimmermann.[13] Wilson's proposal can be stated in the form

$$A(x)B(y) \underset{x \to y}{\to} \sum_{n=1}^{M} C_n(x-y)O_n(y), \qquad (5.88)$$

where A and B are two arbitrary operators, $O_n(y)$ are local operators, and $C_n(x-y)$ are generalized functions which are singular as x and y coincide. Wilson has proposed a conservation of dimension† between the LHS and RHS of 5.88 as a means of estimating the strength of the singularity of $C_n(x-y)$ as $x \to y$. If the 'dimensions' of A, B and O_n are d_A, d_B and d_{O_n} respectively, and s_n is the strength of the singularity of C_n then we have

$$d_A + d_B = s_n + d_{O_n}. \qquad (5.89)$$

For negative s_n the function $C_n(x)$ is singular at $x = 0$. When $n > M$, s_n is positive and the singular part of the asymptotic expansion together with any non-zero finite part terminates. If one arranges the order of the operators O_n such that $d_{O_n} > d_{O_m}$ when $n > m$ then the first term in 5.88 is the most singular and so on. We only study the details of one example. This is the case when $A = B = \phi(x)$. It can be shown[13] in perturbation theory that

$$\phi(x)\phi(y) \underset{x \to y}{\to} C(x-y):\phi^2\left(\frac{x+y}{2}\right): + \cdots \qquad (5.90)$$

where the function $C(x-y)$ is given by

$$C(x) = \tfrac{1}{2}\langle 0|T(\phi(x)\phi(-x)\hat{\phi}(0)\hat{\phi}(0))|0\rangle_c \hat{\Delta}_F^{-2}(0). \qquad (5.91)$$

In 5.91 the suffix C as usual stands for connected part and $\hat{\phi}(p)$ is the Fourier transform of $\phi(x)$ and $\hat{\Delta}_F(p)$ is the Fourier transform of

† This dimension is not quite the same as the ordinary fixed dimension in units of length of a physical quantity. This dimension is the so called scale dimension of an operator. It is obtained by calculating the behaviour of the operator under scale transformations. In general it need not have exactly the same numerical value as the ordinary dimension cf. ref. 6.

$\Delta_F(x)$. These Fourier transforms were defined in Chapter 1. We leave this formula as it stands for the moment and return to the Green's function $\Delta\Gamma(p, -0, p, 0)$ of the last section. This Green's function is the one particle irreducible part of the expression

$$\frac{-im^2}{2(2\pi)^8} \int dx\, dx_1 e^{-ipx_1} \langle 0|T(\phi(x_1)\phi(-x_1):\phi^2(x):\hat{\phi}(0)$$
$$\cdot \hat{\phi}(0))|0\rangle_C \hat{\Delta}_F^{-2}(p)\hat{\Delta}_F^{-2}(0). \tag{5.92}$$

We require the behaviour of this expression for large p. We calculate the behaviour of 5.92 by use[13] of the Riemann–Lebesgue lemma. At large p the rapidly oscillating exponential term should cause the integral to be dominated by the value of the rest of the integrand evaluated at small x_1. In other words by the behaviour of the product $\phi(x_1)$, $\phi(-x_1)$ for small x_1. Now we may insert the Wilson expansion given in 5.90, 91. We then obtain,

$$\frac{-im^2}{4(2\pi)^8} \int dx\, dx_1\, e^{-ipx_1}[\langle 0|T(\phi(x_1)\phi(-x_1)\hat{\phi}(0)\hat{\phi}(0))|0\rangle_C \hat{\Delta}_F^{-2}(0)\hat{\Delta}_F^{-2}(p)]$$
$$\cdot [\langle 0|T(:\phi^2(x)::\phi^2(0):\hat{\phi}(0)\hat{\phi}(0))|0\rangle_C \hat{\Delta}_F^{-2}(0)]. \tag{5.93}$$

We require the one particle irreducible contribution to 5.93. Inspection now shows that the first factor in the integrand of 5.93 is $\Gamma(p, 0, p, 0)$; comparison with 5.75 shows that the second factor is just $\zeta(\lambda)$, i.e. the zero momentum Green's function depicted in Fig. 6. This completes our justification of the use in the last section of the Wilson expansion to calculate asymptotic behaviours at exceptional momenta. Calculations at exceptional momenta provide the closest contact with the actual kinematics of experiments. For more details of such calculations in $\lambda\phi^4/4!$ theory cf. Symanzik.[14] The connection between the Bjorken scaling of the Compton amplitude and perturbation theory is a problem in exceptional momenta treated by Christ et al.[15] The scaling properties of other current amplitudes may be treated in similar fashion, see Mason[16] and Nash.[17]

Infrared and Ultraviolet Stability, Fixed Points

We have noted that the function $\lambda(\mu)$ is an increasing function of μ due to the positivity of $\beta(\lambda)$. In particular for small λ we integrated the equation

$$\mu^2 \frac{d\lambda(\mu)}{d\mu^2} = \beta(\lambda(\mu)) = \frac{3}{2^5\pi^2}\lambda^2(\mu) + \cdots \qquad (5.94)$$

to obtain the approximation

$$\lambda(\mu) = \frac{\lambda}{(1-(3/2^5\pi^2)\lambda \ln \mu^2)} = 1 + \frac{3}{2^5\pi^2}\lambda \ln \mu^2 + \cdots . \qquad (5.95)$$

However this approximation cannot be reliable for value of μ such that $(3\lambda/2^5\pi^2)\ln\mu^2 \gg 1$. In other words these formulae are invalid if we take μ to be very large. On the other hand the same reasoning in reverse allows us to say that $\lambda(\mu)$ decreases as μ goes to zero rather than infinity. In fact if μ goes to zero then inspection of 5.95 shows that $\lambda(\mu)$ goes to zero. This in its turn justifies the truncation of the series for $\beta(\lambda(\mu))$, because $\lambda(\mu)$ can be as small as is needed by choosing μ small enough. The result of this line of reasoning is to convince us that computations of the infrared behaviour of Green's functions, ($\Gamma(\mu p_1, \ldots \mu p_2, m^2, \lambda)$ at small μ), becomes arbitrarily accurate for small enough μ. Further this whole argument has as its basis the positivity of the constant b_0. Field theories with a positive function $\beta(\lambda)$ have been called infrared stable by Wilson.[18] This refers to the infrared computability discussed above. When the function $\beta(\lambda)$ is negative the theory is called ultraviolet stable. We show the graph of $\beta(\lambda)$ in $\lambda\phi^4/4!$ theory for small λ in Fig. 7. It is of

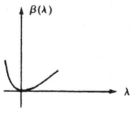

FIG. 7

great interest to know the behaviour of the graph of $\beta(\lambda)$ for larger values of λ. In particular if $\beta(\lambda)$ crosses the λ axis and changes sign, or simply has a zero, there are important consequences. Two possible behaviours for $\beta(\lambda)$ are shown in Figs 8(a) and (b). In Fig. 8(a) β is

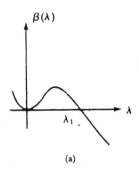
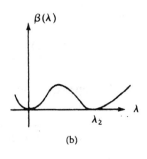

Fig. 8

shown crossing the axis at the point λ_1 and changing sign, in Fig. 8(b), β has a zero at $\lambda = \lambda_2$, but does not change sign. We discuss the graph of Fig. 8(a) first. The graph of Fig. 8(a) gives rise to three possible theories. These three theories are characterized by the values of λ given below

(i) $\qquad\qquad\qquad 0 < \lambda < \lambda_1,$

(ii) $\qquad\qquad\qquad \lambda = \lambda_1,$ $\qquad\qquad\qquad$ (5.96)

(iii) $\qquad\qquad\qquad \lambda > \lambda_1.$

In case (i) of 5.96, $\lambda(\mu)$ decreases to zero as $\mu \to 0$, (infrared stability). If $\mu \to \infty$, $\lambda(\mu)$ increases and approaches the value λ_1 asymptotically. If $\lambda_1 \gg 1$, then the perturbation theory is inapplicable before $\lambda(\mu)$ approaches this value yielding the result that the ultraviolet behaviour is uncomputable by this method. If $\lambda_1 < 1$ it is conceivable that the ultraviolet behaviour may be computable.

Turning now to case (ii) of 5.96; we have $\beta(\lambda_1) = 0$ thus $\lambda(\mu)$ satisfies the equation

$$\mu^2 \frac{d\lambda(\mu)}{d\mu^2} = 0, \qquad \lambda(\mu) = \lambda_1 \qquad \mu = 1 \qquad (5.97)$$

i.e. $\lambda(\mu) = \lambda_1$ for all μ. For this theory $\lambda(\mu)$ remains fixed at the value λ_1 and the asymptotic formula given in 5.56 has to be re-examined. We now do this. Equation 5.56 may be written as

$$\Gamma_{as}(\mu p_1, \ldots \mu p_n, m^2, \lambda)$$
$$= \mu^{4-n} \exp\left[-n \int_\lambda^{\lambda_\mu} dz \frac{\gamma(\lambda_1)}{\beta(z)} - n \int_\lambda^{\lambda(\mu)} \frac{(\gamma(z) - \gamma(\lambda_1))}{\beta(z)} dz\right]$$
$$\cdot \Gamma_{as}(p_1, \ldots p_n, m^2, \lambda(\mu)) \quad (5.98)$$
$$= \mu^{4-n} \exp\left[-n\gamma(\lambda_1) \ln \mu^2\right] \exp\left[-n \int_\lambda^{\lambda(\mu)} dz \frac{(\gamma(z) - \gamma(\lambda_1))}{\beta(z)}\right]$$
$$\cdot \Gamma_{as}(p_1, \ldots p_n, m^2, \lambda(\mu)) \quad (5.99)$$

where we have used 5.52. Next we set $\lambda = \lambda_1$ and allow μ to increase to infinity. The last integral in 5.99 is then zero since $\lambda(\mu) = \lambda = \lambda_1$ and 5.99 becomes

$$\Gamma_{as}(\mu p_1, \ldots \mu p_n, m^2, \lambda_1) = (\mu)^{4-n-2n\gamma(\lambda_1)} \Gamma_{as}(p_1, \ldots p_n, m^2, \lambda_1). \quad (5.100)$$

Furthermore this formula appears to hold for finite μ rather than $\mu = 0$ or ∞. Formula 5.100 which describes the behaviour of Γ_{as} under the scaling $p \to \mu p$ is said to exhibit the scale invariance of a field theory which satisfies $\beta(\lambda_1) = 0$. The parameter $4 - n - 2n\gamma(\lambda_1)$ which differs from the ordinary dimension by the amount $2n\gamma(\lambda_1)$ is said to be the anomalous dimension of $\Gamma_{as}(p_1, \ldots p_n, m^2, \lambda_1)$. The state of affairs which is produced by a zero of β is clearly an exciting possibility. Unfortunately, due to the lack of knowledge of the power series which defines $\beta(\lambda)$, the question of whether β has any zeros is completely open.‡ The theory with a zero is also referred to as a fixed-point theory since the function $\lambda(\mu)$ does not change as μ changes.

Finally, in case (iii) $\lambda > \lambda_1$ and $\beta(\lambda)$ is negative. The function $\lambda(\mu)$ is now a decreasing function of μ, as μ increases $\lambda(\mu)$ tends towards the fixed point λ_1; but as μ decreases, $\lambda(\mu)$ increases to infinity. We have a situation which is exactly complementary to that of case (i). The theory is ultraviolet stable rather than infrared stable.

These three cases correspond to three distinct theories as is clear from the above discussion. A fixed point or zero of $\beta(\lambda)$ may occur without β changing sign. This possibility is shown in Fig. 8(b) where the graph for β just touches the λ axis at the point λ_2. Again we have

‡ The next coefficient b_1 has been calculated. It is negative and equal to $-17/2^9 \cdot 3\pi^4$.[19]

three theories corresponding to

(i) $\qquad 0 < \lambda < \lambda_2,$

(ii) $\qquad \lambda = \lambda_2,$ (5.101)

(iii) $\qquad \lambda_2 < \lambda < \infty.$

However in this case both (i) and (iii) correspond to infrared stable theories because β remains positive in these regions. The fixed-point theory at $\lambda = \lambda_2$ is much the same as the previous one. The fact that the derivative of β also vanishes at $\lambda = \lambda_2$ will affect the form of the corrections to the asymptotic behaviour though. It should be added that $\beta(\lambda)$ may not have a zero at all; it may remain strictly positive for

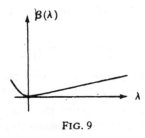

FIG. 9

all λ. We show this in Fig. 9. In this case there is only one theory, the infrared stable one with $\lambda > 0$. We have scrupulously avoided the discussion of theories with $\lambda < 0$. There may seem to be no reason why, after renormalization the coupling constant cannot take on a negative value. Indeed if this were so we would have yet another possible theory in which $\lambda(\mu) \to 0$ as $\mu \to \infty$ and the ultraviolet behaviour would be computable to arbitrary accuracy. However there is an objection that if the interaction term $\lambda \phi^4/4!$ changes sign the energy may not have a lower bound cf. reference 20.

Asymptotic Freedom

It is clear that in perturbation theories $\beta(\lambda)$ is zero for $\lambda = 0$. But if β remains positive for small positive λ then the fixed point at the origin is an infrared stable fixed point. In order to compute accurately the ultraviolet or large-momentum behaviour of Green's functions we require the origin to be an ultraviolet-stable fixed point. This requires β to be negative for small positive λ. The class of renormalizable perturbation theories for which the origin is an ultraviolet-stable fixed point is known.[20] (This class has to be extended if one considers model field theories for which the dimension of space time is other than 4). Such theories have been named asymptotically free theories.[20] This is because $\lambda(\mu)$ tends to the fixed point zero asymptotically when $\mu \to \infty$, i.e. the theory is asymptotic to the free theory. This means that the Born term or first few terms of the perturbation series should dominate in the large momentum limit. This is obviously a most attractive state of affairs. The examples of asymptotically free theories must contain the Yang–Mills field.[20] Since we have not yet met the Yang–Mills field, we shall now introduce it and compute its β function near the origin.

The theory of the Yang–Mills field is a generalization of the electromagnetic field to a case where the gauge group is a non-Abelian group, say SU(2), rather than the Abelian U(1), as it is in QED. The Lagrangian for the Yang–Mills field is

$$L = -\tfrac{1}{4} F^a_{\mu\nu} F^{a\mu\nu} - \frac{1}{2\xi}(\partial_\lambda A^{a\lambda})^2 - \bar{c}^a \partial^\mu D_\mu^{ab} c^b. \qquad (5.102)$$

The notations and definitions used in 5.102 are as follows: $A^a_\mu(x)$ is the Yang–Mills field, the analogue of the photon field A_μ. The index a indicates that it is a vector with respect to the transformation of the non-Abelian gauge group, (SU(2)). The field tensor $F^a_{\mu\nu}$ is defined by

$$F^a_{\mu\nu} = \partial_\mu A^a_\nu(x) - \partial_\nu A^a_\mu(x) + ig T^a_{bc} A^b_\mu(x) A^c_\nu(x) \qquad (5.103)$$

T^a_{bc} is a matrix element. The index a indicates which matrix we are dealing with and the labels b and c indicate the bcth entry in the matrix for T^a. T^a is a Hermitean matrix whose matrix elements are directly proportional to the structure constants of the group. In fact we have

$$T^a_{bc} = -i f_{abc}. \qquad (5.104)$$

This means that the matrices T^a carry a representation of the Lie algebra of the gauge group for we can verify that

$$[T^a, T^b] = i f_{abc} T^c. \qquad (5.105)$$

It is simplest to have in mind SU(2). For SU(2), a has the values 1, 2, 3 and $f_{abc} = \varepsilon_{abc}$. (We could, of course, consider other representations of SU(2) not just the one defined by 5.104.) The structure of the infinitesimal gauge transformations is altered in the non-Abelian case as compared with the Abelian case of QED. The term $F^a_{\mu\nu} F^{a\mu\nu}$ is invariant under the gauge transformation

$$A_\mu^a(x) \to A_\mu^a(x) + D_\mu^{ab}\,\delta\Lambda^b(x) \qquad (5.106)$$

where $\delta\Lambda^b(x)$ is an infinitesimal gauge parameter and D_μ^{ab} is the covariant derivative given by

$$D_\mu^{ab} = \partial_\mu\,\delta^{ab} - ig\mathrm{T}^c_{ab}A_\mu^c(x) \qquad (5.107)$$

(in QED T = 0 so $D_\mu = \partial_\mu$). To construct the Feynman rules one adds to the Lagrangian the usual gauge-fixing term $(\partial_\lambda A^{\lambda a})^2/-2\xi$. We then refer to the first section of Chapter 4 which examines the consequences of the gauge transformation properties of the gauge fixing term. In general we saw that a ghost term must be added to the Lagrangian. It takes a form given by eq. 4.10 of Chapter 4. In QED we saw that this ghost term corresponded to a free field and so had no consequences. (We are not now considering the gauge fixing term of the third section of Chapter 4, in which QED is formulated with ghosts.) The Yang–Mills theory does have an interacting ghost term. This is because of the presence of the extra term linear in A_μ^a in the gauge transformation 5.106. The form of this ghost term may be easily calculated using eq. 4.10 of Chapter 4. This results in the ghost term shown in the Lagrangian of 5.102. The Feynman rules that result from this Lagrangian are

vector propagator: $\mu \sim\!\!\sim\!\!\sim\nu$, momentum p

$$-i\delta^{ab}\left[\left(g_{\mu\nu} - \frac{p_\mu p_\nu}{p^2}\right)\frac{1}{p^2} + \frac{\xi p_\mu p_\nu}{p^4}\right]$$

Three-gluon vertex with legs (a,λ,p), (b,μ,q), (c,ν,r), $p+q+r=0$:

$$-ig\mathrm{T}^a_{bc}[(p-q)_\nu g_{\lambda\mu} + (q-r)_\lambda g_{\mu\nu} + (r-p)_\mu g_{\nu\lambda}]$$

Four-gluon vertex with legs (a,λ), (d,σ), (b,μ), (c,ν):

$$ig^2\mathrm{T}^a_{be}\mathrm{T}^c_{de}(g_{\lambda\nu}g_{\mu\sigma} - g_{\lambda\sigma}g_{\mu\nu})$$
$$+ig^2\mathrm{T}^a_{ce}\mathrm{T}^b_{de}(g_{\lambda\mu}g_{\nu\sigma} - g_{\lambda\sigma}g_{\mu\nu})$$
$$+ig^2\mathrm{T}^a_{de}\mathrm{T}^c_{be}(g_{\lambda\nu}g_{\mu\sigma} - g_{\lambda\mu}g_{\sigma\nu})$$

symmetry factor $\;\mathcal{S}$

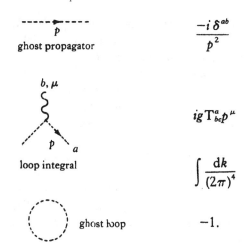

Now we wish to calculate $\beta(g)$ for small g. To do this we require the form of the Callan–Symanzik equation for the Yang–Mills theory. We cannot use the technique of mass-vertex insertion here since we are dealing with a theory in which all the masses are zero. Nevertheless the renormalized Green's functions $\Gamma_{\mu_1...\mu_n}(p_1,...p_n)$ do depend on a parameter M with the dimensions of mass. It is by considering how this parameter M arises, that we can derive the form of the Callan–Symanzik equation for Yang–Mills theories, and for any other zero-mass theory. We denote the unrenormalized Green's functions by $\Gamma^U(p_1...p_n)$. These Green's functions have no dependence on a mass parameter. Nevertheless when we renormalize the theory there is an arbitrariness in the finite part which has to be dealt with. This is dealt with by imposing normalization conditions on the Green's functions, for example, those of eqs 5.19, 20. To be specific let a mass-parameter M enter by the use of the D-dimensional coupling constant $M^{(4-D)/2}g$. This quantity when expanded in $\ln M^2$ combines with the pole terms to give finite parts which we fix by the normalization conditions

$$\Gamma^{ab}_{\mu\nu}(p,-p,M,g) = -i\delta^{ab}[g_{\mu\nu}(-M^2 - p_\mu p_\nu) + \xi^{-1} p_\mu p_\nu]$$
$$p^2 = -M^2 \tag{5.108}$$

$$\Gamma^{abc}_{\lambda\mu\nu}(p,q,r) = -ig T^a_{bc}[-q_\nu g_{\lambda\mu} + 2q_\lambda g_{\mu\nu} - q_\mu g_{\nu\lambda}]$$
$$p = 0, \quad p+q+r = 0, \quad q^2 = r^2 = -M^2. \tag{5.109}$$

In the above $-\Gamma^{ab}_{\mu\nu}$ is the inverse-propagator and $\Gamma^{abc}_{\lambda\mu\nu}$ is the three-point vertex. The normalization point is chosen at the spacelike point $-M^2$ so as to avoid being on the branch cuts of the momentum variables where there are additional physical singularities due to unitarity. These branch cuts are present for all positive momenta p^2 since we are dealing with a zero-mass theory. Also the reason that we have not chosen a symmetric point definition of the normalization condition for $\Gamma^{abc}_{\lambda\mu\nu}$ is to make the Feynman integrals which give b_0 easier to evaluate.

The general properties of the mass-dependence of the theory are that the Green's functions depend explicitly on M^2 through 5.108, 5.109; and that the coupling constant g depends on M^2 through 5.109(b). This in turn means that the renormalization constants Z which depend on g also depends on M^2. If $\Gamma^{a_1...a_n}_{\mu_1...\mu_n}$ is a renormalized Green's function with n external legs we can display the M^2 dependence in the basic renormalization equation

$$[Z(M^2)]^{n/2} \Gamma^{Ua_1...a_n}_{\mu_1...\mu_n}(p_1,\ldots p_n, \xi_0, g_0)$$
$$= \Gamma^{a_1...a_n}_{\mu_1...\mu_n}(p_1 \ldots p_n, M^2, \xi(M^2), g(M^2)). \quad (5.110)$$

Now $\Gamma^{Ua_1...a_n}_{\mu_1...\mu_n}(p_1,\ldots p_n, \xi_0, g_0)$ of 5.110 does not depend on M^2 because it is a Green's function of the unrenormalized theory. Therefore if we differentiate 5.110 with respect to M^2 and apply the chain rule we obtain†

$$\left[M^2 \frac{\partial}{\partial M^2} + \beta(g)\frac{\partial}{\partial g} + \delta(g)\frac{\partial}{\partial \xi} - n\gamma(g)\right] \Gamma^{a_1...a_n}_{\mu_1...\mu_n}(p_1,\ldots p_n, \xi, g, M^2) = 0$$
$$(5.111)$$

where

$$\beta = M^2 \frac{dg}{dM^2},$$
$$\delta = M^2 \frac{d\xi}{dM^2}, \quad (5.112)$$
$$\gamma = \frac{1}{2} M^2 \frac{d\ln Z}{dM^2}.$$

The equation that we required is that given in 5.111 and we see that apart from the addition of the ξ derivative it is obtained by setting the

† The renormalized gauge parameter ξ depends on M^2 through 5.109(a).

ASYMPTOTIC BEHAVIOUR

$\Delta\Gamma$ term of the massive theory to zero. The absence of the $\Delta\Gamma$ term in the Callan–Symanzik equation is a characteristic of massless theories. We are now ready to calculate $\beta(g)$. This proceeds in an exactly analogous way to the calculation in $\lambda\phi^4/4!$ theory. This time we must first calculate $\gamma(g)$. This entails evaluating the equation for $\Gamma^{ab}_{\mu\nu}(p,-p)$ at the normalization point 5.109(a) and then inserting the lowest order graphs shown in Fig. 10. This calculation may be done in more than

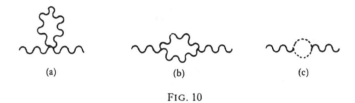

FIG. 10

one gauge. We choose to work with the Landau gauge for which $\xi = 0$. This choice of gauge does affect the expression for $\gamma(g)$ since γ is defined in terms of the gauge dependent quantity $\Gamma^{ab}_{\mu\nu}$ and so depends on ξ. However $\beta(g)$ depends on g which should not depend on ξ; and the value of b_0 is unchanged if one works with $\xi = 1$ for example. Next some remarks about the function $\delta(g)$ are useful. The renormalized gauge parameter ξ is obtained from the renormalized two-point function. It is therefore related to the corresponding bare-parameter ξ_0 by

$$\xi = Z^{-1}\xi_0, \qquad (5.113)$$

hence δ may be written as

$$\delta = -\frac{M^2}{Z}\frac{dZ}{dM^2}\xi = -2\gamma\xi. \qquad (5.114)$$

Still on the subject of the gauge properties of $\Gamma^{ab}_{\mu\nu}$ we point out that the methods of Chapter 4 may be used to derive the consequences of gauge invariance for $\Gamma^{ab}_{\mu\nu}$. The result is a form for $\Gamma^{ab}_{\mu\nu}$ extremely similar to the form for the inverse photon propagator (eq. 4.63 Chapter 4). We find

$$\Gamma^{ab}_{\mu\nu} = -i\,\delta^{ab}[(g_{\mu\nu}p^2 - p_\mu p_\nu)d^{-1} + \xi^{-1}p_\mu p_\nu]. \qquad (5.115)$$

We should warn the reader that though in QED the function d^{-1} is not a function of ξ this is not the case in Yang–Mills theories. If we use

5.114,115 we may write the Callan–Symanzik equation as an equation for d^{-1} only. The result of this is the equation

$$\left[M^2\frac{\partial}{\partial M^2}+\beta\frac{\partial}{\partial g}-2\gamma\xi\frac{\partial}{\partial \xi}-2\gamma\right]d^{-1}(p,M,g,\xi)=0. \quad (5.116)$$

But since d^{-1} is dimensionless we may rewrite this as

$$\left[-p^2\frac{\partial}{\partial p^2}+\beta\frac{\partial}{\partial g}-2\gamma\xi\frac{\partial}{\partial \xi}-2\gamma\right]d^{-1}=0. \quad (5.117)$$

Applying the normalization condition 5.109(a) we obtain the equation

$$2\gamma\left\{1+\xi\frac{\partial d^{-1}(p^2=-M^2)}{\partial \xi}\right\}=M^2\frac{\partial d^{-1}(p^2=-M^2)}{\partial p^2}, \quad (5.118)$$

in the Landau gauge we obtain

$$\gamma(g)=\frac{M^2}{2}\frac{\partial d^{-1}(p^2=-M^2,\xi=0)}{\partial p^2}. \quad (5.119)$$

The next task is the evaluation of the graphs of Fig. 10 and their insertion into 5.119. These graphs are quite easy though time-consuming to evaluate. The properly normalized contribution to d^{-1} that they give is simply

$$C_1\frac{13}{3}\frac{g^2}{2^5\pi^2}\ln\left(\frac{p^2}{-M^2}\right) \quad (5.120)$$

where C_1 is defined by

$$T^a_{ef}T^b_{ef}=-C_1\delta^{ab} \quad (5.121)$$

In this case

$$T^a_{bc}=-if_{abc}=-i\varepsilon_{abc} \quad (5.122)$$

so that $C_1=2$.

Thus if we write

$$\gamma(g)=d_0g^2+d_1g^3+\cdots$$

then 5.119,120 give straightaway that

$$d_0=\frac{-13}{3\cdot 2^5\pi^2}. \quad (5.123)$$

ASYMPTOTIC BEHAVIOUR

We remind the reader that in evaluating the graphs of Fig. 10 the first two have a symmetry factor of $\frac{1}{2}$ and the last has a ghost loop factor of -1. Also we must always choose only the one-particle irreducible graphs.

To calculate β we must use the normalization condition 5.109(b) on the Callan–Symanzik equation for $\Gamma^{abc}_{\lambda\mu\nu}(p, q, r)$. Let us denote the momenta of the normalization point 5.109(b) by $\hat{p}\,\hat{q}\,\hat{r}$, and let us define $\Gamma(p, q, r)$ by

$$\Gamma^{abc}_{\lambda\mu\nu}(p, q, r) = -iT^a_{bc}[(p-q)_\nu g_{\lambda\mu} + (q-r)_\lambda g_{\mu\nu} + (r-p)_\mu g_{\nu\lambda}]$$
$$\cdot \Gamma(p, q, r) + \cdots. \qquad (5.124)$$

The dots in 5.124 stand for terms of other tensor structure. The equation for β then amounts to

$$\beta(g) = b_0 g^3 + b_1 g^4 + \cdots$$
$$= -M^2 \frac{\partial \Gamma(\hat{p}, \hat{q}, \hat{r})}{\partial M^2} + 3\gamma g. \qquad (5.125)$$

To calculate the first term in 5.125 one must evaluate the graphs of Fig. 11 at $\hat{p}, \hat{q}, \hat{r}$. Again these present no technical problem particularly

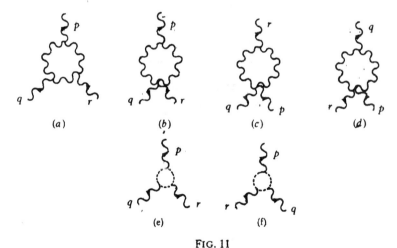

FIG. 11

with the normalization point that we have chosen. Details to remember are symmetry factors of $\frac{1}{2}$ for graphs (b), (c) and (d) and minus signs for graphs (e) and (f). The actual number we obtain for $\Gamma(\hat{p}, \hat{q}, \hat{r})$

is

$$C_1 \frac{17}{3 \cdot 2^6 \pi^2} g^3 \ln\left(\frac{q^2}{-M^2}\right). \tag{5.126}$$

Combining 5.123,125,126 we finally find that

$$b_0 = \frac{-11}{3 \cdot 2^4 \pi^2}. \tag{5.127}$$

We finish this section by giving an example of a calculation with an asymptotically free theory. Consider the inverse propagator $-\Gamma^{ab}_{\mu\nu}(p, -p, M^2; g)$. The large momentum behaviour of $\Gamma^{ab}_{\mu\nu}$ is simply a problem in obtaining the large momentum behaviour of the function d^{-1} defined in 5.115. Using the techniques already derived we can write

$$d^{-1}(\mu p, M, g) = \exp\left[-2 \int_g^{g(u)} dz \frac{\gamma(z)}{\beta(z)}\right] d^{-1}(p, M, g(u)). \tag{5.128}$$

In this case $g(u)$ is obtained by solving

$$\mu^2 \frac{dg(u)}{d\mu^2} = \beta(g(u)) = -\frac{11}{3 \cdot 2^4 \pi^2} g^3(u) + \cdots. \tag{5.129}$$

Also we know that from perturbation theory

$$d^{-1}(p, M, g(u)) = 1 + O(g(u)), \tag{5.130}$$

the asymptotic form d_{as}^{-1} for large μ is then

$$d_{as}^{-1} = \exp\left[-2 \int_g^{g(u)} \frac{dz}{z} \frac{d_0}{b_0}\right][1 + O(g(u))] \tag{5.131}$$

i.e.

$$d_{as}^{-1} = \left[\frac{g(u)}{g}\right]^{-2d_0/b_0} = \left[\frac{1}{(1 - 2b_0 g^2 \ln \mu^2)^{1/2}}\right]^{-2d_0/b_0}$$

$$= \left(1 + \frac{11}{3 \cdot 2^3 \pi^2} g^2 \ln \mu^2\right)^{13/22} \tag{5.132}$$

Further we have the good fortune that this approximation improves as μ increases and $g(u)$ goes to zero. For details of the applications of asymptotically free theories to Bjorken scaling and strong interaction problems see Gross and Wilczek.[21]

The Callan–Symanzik Equations for QED

We have not yet mentioned the Callan–Symanzik equations in connection with QED. They are simple to derive and take the form that one would expect.[22] The mass insertion operator is $\bar{\psi}(x)\psi(x)$, the photon being massless. If $\gamma(e)$ is the anomalous dimension of the Fermion and $\gamma_3(e)$ the anomalous dimension of the photon the equation takes the form

$$\left[m^2 \frac{\partial}{\partial m^2} + \beta(e)\frac{\partial}{\partial e} - 2\gamma_3 \xi \frac{\partial}{\partial \xi} - 2n\gamma - m\gamma_3\right]$$

$$\Gamma_{\mu_1 \ldots \mu_m}(p_1, \ldots p_n, q_1, \ldots q_n, r_1, \ldots r_m) = \Delta\Gamma_{\mu_1 \ldots \mu_m} \quad (5.133)$$

where $\Gamma_{\mu_1 \ldots \mu_m}(p_1, \ldots p_n, q_1, \ldots q_n, r_1, \ldots r_m)$ is a one-particle irreducible Green's function with $2n$ Fermion legs and m photon legs. $\Delta\Gamma_{\mu_1 \ldots \mu_m}$ is the corresponding Green's function with the Fermion mass-operator insertion.

The Connection between Zero Mass Theories and Γ_{as}

We have seen in the last section that the Callan–Symanzik equations for a theory with zero mass take on the homogeneous form

$$\left[M^2 \frac{\partial}{\partial M^2} + \beta \frac{\partial}{\partial \lambda} - n\gamma\right]\Gamma = 0. \quad (5.134)$$

This is precisely the equation that we solved in the massive case to obtain the asymptotic form which we called Γ_{as}. The question arises as to what connection is there between Γ(massless), the Green's functions of the zero mass theory, and the Γ_{as} of the massive theory? Well the functions Γ(massless) satisfy the zero mass normalization conditions for $\lambda\phi^4/4!$ theory

$$\Gamma(\text{massless}, p, -p, M^2, \lambda') = 0, \qquad p = 0;$$
$$\Gamma(\text{massless}, p, -p, M^2, \lambda') = -iM^2, \qquad p^2 = -M^2; \quad (5.135)$$
$$\Gamma(\text{massless}, p_1, p_2, p_3, p_4) = -i\lambda', \qquad p_i \cdot p_j = -\tfrac{1}{3}(4\delta_{ij} - 1)M^2.$$

Apart from this Γ_{as} and Γ(massless) obey the same form of partial differential equation. What this amounts to is that Γ_{as} and Γ(massless) differ only by a finite renormalization. Let the finite renormalization constant that accomplishes this be $Z(\lambda)$. Then the equation which relates Γ_{as} and $\Gamma(\text{massless})$ is

$$\Gamma_{as}(p_1, \ldots p_n, m^2, \lambda) = [Z(\lambda)]^{-n/2}\Gamma(\text{massless}, p_1, \ldots p_n, M^2, \lambda'(\lambda)) \quad (5.136)$$

where we have chosen the normalization mass M to be m for convenience. The constant $Z(\lambda)$ is a power series in λ starting with unity. $Z(\lambda)$ may be calculated by imposing the normalization conditions 5.135 and using eq. 5.136. The content of this discussion is that Γ_{as} may, via $Z(\lambda)$, be constructed directly by writing down the perturbation theory graphs for a zero mass $\lambda\phi^4/4!$ theory.

The 't Hooft–Weinberg Equations

We now come to our second set of renormalization group equations.[23] Let us consider the vertex function $\Gamma(p_1, \ldots p_n, m^2, \lambda)$ of massive $\lambda\phi^4/4!$ theory. Next we allow the mass parameter M^2 to enter into the theory in addition to the renormalized mass m^2. We do this by using the coupling constant $\lambda(M^2)^{(D-4)/2}$ during the regularization procedure. This means that the renormalization constant Z, the renormalized mass and coupling will all be functions of M^2. We display this by writing $\Gamma(p_1, \ldots p_n, m^2, \lambda)$ as

$$\Gamma \equiv \Gamma(p_1, \ldots p_n, m^2, M^2, m^2(M^2), \lambda(M^2)). \quad (5.137)$$

Now let us define the (infinite) constant Z_m that relates the bare to the renormalized mass by

$$m^2 = Z_m m_0^2 \quad (5.138)$$

(m_0 is the bare mass). Normally we expect that this constant depends on m as well as the coupling constant so that we would write

$$Z_m \equiv Z_m(\lambda, m^2). \quad (5.139)$$

Also Z_m is a complicated function of λ and m^2. Therefore eq. 5.138 is really a complicated and non-linear relation for m^2 with m^2 appearing on both sides of the equation. However, the particular renormalization procedure used in this section is so constructed that the constant Z_m is not a function of m^2. (It is always a function of λ, of course). A simple line of argument can be used to establish this. In the dimensional method the infinities that arise are simple or multiple poles in D at

$D = 4$. We formalize this by eq. 2.70 of Chapter 2 with slightly modified notation

$$m_0^2 = Z_m^{-1} m^2 = \sum_{i=0}^{\infty} \frac{d_i(M^2, \lambda, m^2)}{(D-4)^i}. \quad (5.140)$$

Now M^2 only occurs as the quantity $(M^2)^{(D-4)/2}$; this allows terms in $\ln M^2$ to appear when the poles are cancelled by the regularization procedure. Further the residue of the poles contain the parameter m^2 which appears always in a polynomial, i.e. as a positive power. Finally the d_i have dimension (mass)^2. This reasoning means that the d_i cannot depend on M^2 and preserve the above properties. The dimension of d_i thus means that it may be written as

$$d_i = \hat{d}_i m^2 \quad (5.141)$$

where $\hat{d}_i$ depends neither on m^2 nor, a fortiori, on M^2. Insertion of 5.141 into 5.140 means that Z_m^{-1} depends only on λ as required. The same reasoning can be employed to show that the coupling constant renormalization and the field renormalization constant have no mass dependence. A general discussion of these renormalization prescriptions and a source of useful examples of the properties of the 't Hooft–Weinberg equations is given by Collins and McFarlane.[24]

The renormalization of $\Gamma(p_1, \ldots p_n, m^2, \lambda)$ is expressed by the equation

$$\Gamma^U(p_1, \ldots p_n, m_0^2, \lambda_0) = Z^{-n/2}(\lambda) \Gamma(p_1, \ldots p_n, \lambda, M^2, m^2) \quad (5.142)$$

with Γ^U the renormalized Green's function. But Γ^U does not depend on the renormalization parameter M^2. Therefore if we differentiate with respect to M^2 we obtain

$$\left[M^2 \frac{\partial}{\partial M^2} + \beta \frac{\partial}{\partial \lambda} + \gamma_\theta m^2 \frac{\partial}{\partial m^2} - n\gamma \right] \Gamma(p_1, \ldots p_n, \lambda, M^2, m^2) = 0 \quad (5.143)$$

where

$$\beta(\lambda) = M^2 \frac{d\lambda}{dM^2}$$

$$\gamma(\lambda) = \frac{1}{2} M^2 \frac{d \ln Z}{dM^2} \quad (5.144)$$

$$\gamma_\theta(\lambda) = \frac{M^2}{m^2} \frac{dm^2}{dM^2} = M^2 \frac{d \ln Z_m}{dM^2}.$$

We have ensured that the functions of 5.144 depend only on $\lambda(M)$ and not on M^2/m^2 as well by our mass independent renormalization procedure. The first advantage that we see in 5.143 as compared with the Callan–Symanzik equation is that it is a homogeneous equation. The term $\gamma_\theta m^2(\partial/\partial m^2)$ has replaced the term $\Delta\Gamma$. It can easily be verified by differentiation that

$$\Gamma(p_1,\ldots p_n, m^2, M^2, \lambda) = \exp\left[n\int_{\lambda_0}^{\lambda} dz \frac{\gamma}{\beta}\right]$$
$$\cdot f^{p_1\cdots p_n}\left(\ln M^2 - \int_{\lambda_0}^{\lambda} \frac{dz}{\beta}, \ln m^2 - \int_{\lambda_0}^{\lambda} \frac{\gamma_\theta}{\beta} dz\right) \quad (5.145)$$

satisfies 5.143 for any function $f(x, y)$. Now we return to the dimensional identity

$$\Gamma(\mu p_1, \ldots \mu p_n, m^2, M^2, \lambda) = \mu^{4-n} \Gamma\left(p_1, \ldots p_n, \frac{m^2}{\mu^2}, \frac{M^2}{\mu^2}, \lambda\right). \quad (5.146)$$

Thus

$$\Gamma(\mu p_1, \ldots \mu p_n, m^2, M^2, \lambda) = \mu^{4-n} \exp\left[\int_{\lambda_0}^{\lambda} n\, dz \frac{\gamma}{\beta}\right]$$
$$\cdot f^{p_1\cdots p_n}\left(\ln \frac{M^2}{\mu^2} - \int_{\lambda_0}^{\lambda} \frac{dz}{\beta}, \ln \frac{m^2}{\mu^2} - \int_{\lambda_0}^{\lambda} dz \frac{\gamma_\theta}{\beta}\right). \quad (5.147)$$

We have already seen the value in eq. 5.54 of using the effective coupling constant $\lambda(\mu)$. The form of 5.147 suggests that we should try to absorb the μ^2 dependence in the second argument of the function f. This can be done by introducing the effective mass parameter $m^2(\mu)$ defined by

$$\mu^2 \frac{dm^2(\mu)}{d\mu^2} = [-1 + \gamma_\theta(\lambda(\mu))] m^2(\mu), \quad m^2(\mu) = m^2, \quad \mu = 1. \quad (5.148)$$
$$\mu = 1.$$

This is the differential form of the integrated equation (which, of course, led us to the defining equation (5.148)),

$$\ln \frac{m^2}{\mu^2} - \int_{\lambda_0}^{\lambda} dz \frac{\gamma_\theta}{\beta} = \ln m^2(\mu) - \int_{\lambda_0}^{\lambda(\mu)} dz \frac{\gamma_\theta}{\beta}$$

i.e.

$$\ln\left[\frac{m^2(\mu)}{m^2}\mu^2\right] = \int_\lambda^{\lambda(\mu)} dz \frac{\gamma_\theta}{\beta}$$

or

$$m^2(\mu) = \frac{m^2}{\mu^2}\exp\left[\int_\lambda^{\lambda(\mu)} dz \frac{\gamma_\theta(z)}{\beta(z)}\right]. \tag{5.149}$$

Using $\lambda(\mu)$ and $m^2(\mu)$, the dimensional relation for Γ takes on the extremely compact and simple form

$$\Gamma(\mu p_1, \ldots \mu p_n, m^2, M^2, \lambda) = \mu^{n-4}\exp\left[-n\int_\lambda^{\lambda(\mu)} dz \frac{\gamma}{\beta}\right]$$
$$\cdot \Gamma(p_1, \ldots p_n, m^2(\mu), M^2, \lambda(\mu)). \tag{5.150}$$

We wish to obtain asymptotic forms from formula 5.150. This is done by taking μ to be very large on the LHS. On the RHS the μ dependence resides wholly in $m^2(\mu)$ and $\lambda(\mu)$. We have already analysed the μ dependence of $\lambda(\mu)$. The μ dependence of $m^2(\mu)$ hinges on the observation that if, in eq. 5.148, $[-1+\gamma_\theta(\lambda,\mu)]$ is negative, then $m^2(\mu)$ is a decreasing function of μ, decreasing to zero. The function $\gamma_\theta(\lambda)$ has a power series expansion in λ, for $\lambda\phi^4/4!$ theory

$$\gamma_\theta = f_0\lambda + f_1\lambda^2 + \cdots. \tag{5.151}$$

In QED where the same equations apply it can be verified that

$$\gamma_\theta(e^2) = -\frac{3e^2}{16\pi^2} + \cdots, \tag{5.152}$$

thus to this order γ_θ is negative, so therefore is $[-1+\gamma_\theta]$, and so $m^2(\mu)$ decreases. If we assume that $m^2(\mu)$ decreases we may make a Taylor expansion of Γ in $m^2(\mu)$

$$\Gamma(p_1, \ldots p_n, m^2(\mu), M^2, \lambda(\mu))$$
$$= \Gamma(p_1, \ldots p_n, 0, M^2, \lambda(\mu))$$
$$+ m^2(\mu)\frac{\partial\Gamma}{\partial m^2(\mu)}(p_1, \ldots p_n, 0, M^2, \lambda(\mu))$$
$$+ \frac{m^4(\mu)}{2!}\frac{\partial^2\Gamma}{\partial m^4(\mu)}(p_1, \ldots p_n, 0, M^2, \lambda(\mu)) + \cdots. \tag{5.153}$$

The first term in 5.153 is the vertex function of a zero-mass theory. If we insert this term into formula 5.150 we obtain an asymptotic form Γ_{as} for Γ at large μ. But as we argued in the preceding section the vertex functions of the zero-mass theory are, apart from a finite renormalization, the asymptotic forms Γ_{as} obtained by solving the Callan–Symanzik equation with $\Delta\Gamma$ set equal to zero. Thus the first term in 5.153 is simply the asymptotic form Γ_{as}, already obtained with the Callan–Symanzik equations. The other terms in the Taylor expansion 5.153 are, since $m^2(\mu)$ is decreasing, corrections to the asymptotic form Γ_{as}. Caution should be exercised here though. This Taylor expansion cannot be continued beyond the three terms displayed in 5.153. That is to say only two correction terms to the asymptotic form Γ_{as} may be obtained by this method. The reason for this is that the higher derivatives contain infrared divergences which prevent these derivatives from existing. This is based on the observation, that perturbation theory renormalized in this fashion can typically generate a dependence on m of form $m^3 \ln m^2 = f$. This function f can only be expanded about $m = 0$ as far as the second derivative term in a Taylor series. In QED, for example, let us consider an internal electron line of momentum k in an arbitrary loop. We may write the contribution from this loop in the form $\int (dk/(\not{k} - m)) f(k, m, P)$ for some function f and where P stands for the other momenta. It is easy to convince oneself by drawing diagrams that f is finite when k and m are zero. If we integrate this expression in the neighbourhood of $k = 0$ then, since f is finite, any singularities that are present will appear by simply evaluating the quantity $\int (dk/\not{k} - m)$ for small k. We can verify straight away that $\int (dk/\not{k} - m)$ is finite and twice differentiable at $m = 0$ but that the third derivative diverges logarithmically at $m = 0$. This argument yields the general singularity to be of the form $m^3 \ln m^2$ or a power of this quantity. However this is only true when the momenta of the external lines are non-exceptional. For exceptional momenta the external momenta are such that zero momentum can be fed in at various points on an internal loop. This results in the function $f(k, m, P)$ no longer being finite when k and m are zero. For now two or more internal charged lines may have the same momentum and we must examine expressions like $\int (dk/(\not{k} - m)^2)$ and $\int (dk/(\not{k} - m)^4)$ etc. when expanded about $m = 0$. The Taylor expansion is thus seen in general to break down for exceptional momenta.

Nevertheless we still obtain two correction terms to the asymptotic form Γ_{as}. However these correction terms will also cease to exist if the momenta are exceptional since in this case, as we have seen already, the infrared problem increases. For exceptional momenta the expansion 5.153 cannot be made and some other method must be used to obtain the true asymptotic form Γ_{as}^E. In fact we have already shown that the technique needed to deal with exceptional momenta and to obtain Γ_{as}^E is the Wilson short distance expansion. Finally a remark about $m^2(\mu)$ in asymptotically free theories. In such theories since $\lambda(\mu) \to 0$ for large μ, eq. 5.148 implies that $m^2(\mu)$ is controlled solely by the first term (f_0) in the expansion for γ_θ. Thus the verification that γ_θ is negative to this order is, in the case of asymptotically free theories, sufficient to conclude that $m^2(\mu)$ tends to zero for large μ.

Non-leading Asymptotic Behaviour from the Callan–Symanzik Equations

In the last section we showed that, for non-exceptional momenta, we can use the 't Hooft–Weinberg equations to derive correction terms to the asymptotic form Γ_{as}. This is also possible with the Callan–Symanzik equations. We begin with the Callan–Symanzik equation for $\Gamma(p_1, \ldots, p_n, m^2, \lambda)$ with non-exceptional $p_1 \ldots p_n$,

$$D\Gamma(p_1, \ldots p_n, m^2, \lambda) = \Delta\Gamma. \tag{5.154}$$

To obtain corrections to Γ_{as} we require information about the asymptotic behaviour of $\Delta\Gamma$. This can be done by realizing that $\Delta\Gamma$ is itself a Green's function and therefore obeys its own Callan–Symanzik equation namely

$$[D - \gamma_{\phi^2}(\lambda)]\Delta\Gamma = \Delta(\Delta\Gamma). \tag{5.155}$$

The notation $\Delta(\Delta\Gamma)$ means the Green's function Γ with two mass insertions rather than one. The function $\gamma_{\phi^2}(\lambda)$ is the anomalous dimension of the ϕ^2 mass insertion operator. That is to say it is the derivative of the logarithm of the renormalization constant that renormalizes ϕ^2. So γ_{ϕ^2} bears the same relation to ϕ^2 as the anomalous dimension γ does to the field ϕ. In fact γ_{ϕ^2} is the same as the functions $\gamma_\phi(\lambda)$ or $\zeta(\lambda)$ apart from a differing renormalization prescription. We obtain $\Delta\Gamma_{as}$ by solving

$$[D - \gamma_{\phi^2}]\Delta\Gamma_{as} = 0 \tag{5.156}$$

(since $\Delta(\Delta\Gamma)$ is asymptotically negligible). According to our established techniques this leads at once to the formula

$$\Delta\Gamma_{as}(\mu p_1, \ldots \mu p_n, m^2, \lambda) = \mu^{4-n} \exp\left[\int_\lambda^{\lambda(\mu)} dz \frac{(-n\gamma - \gamma_{\phi^2})}{\beta}\right]$$
$$\cdot \Delta\Gamma_{as}(p_1, \ldots p_n, m^2, \lambda(\mu)). \quad (5.157)$$

The once-corrected asymptotic form of Γ which we write as $\Gamma_{as}^{(1)}$ is obtained by inserting the formula for $\Delta\Gamma_{as}$ in the general solution 5.67. The expression then obtained can be expanded in the usual way to obtain the leading behaviour of the correction term. Formally we have

$$\Gamma_{as}^{(1)}\left(p_1, \ldots p_n, \frac{m^2}{\mu^2}, \lambda\right)$$
$$= \Gamma_{as}\left(p_1, \ldots p_n, \frac{m^2}{\mu^2}, \lambda\right) + \int_{\mu^2}^\infty \frac{d\mu'^2}{\mu'^2} \exp\left[n\int_{\lambda(\mu/\mu')}^\lambda dz \frac{\gamma}{\beta}\right]$$
$$\cdot \Delta\Gamma_{as}\left(p_1, \ldots p_n, \frac{m^2}{\mu'^2}, \lambda\left(\frac{\mu}{\mu'}\right)\right). \quad (5.158)$$

with $\Delta\Gamma_{as}$ given by 5.157. An example of this calculation carried out in detail is given in ref. 14. Generalizations of this procedure using Wilson expansions can be used to obtain still more correction terms but the expressions are complicated to handle.

Minkowskian versus Euclidean Momenta

Throughout these discussions on asymptotic behaviours we have appealed for the use of Euclidean momenta. For Euclidean momenta we restrict the four momenta so that they are always spacelike. We can achieve this by using four momenta with three real components and one imaginary component. There is a reason why we believe that we can make physical sense out of these Euclidean results. It is that we have some prejudice about the analytic behaviour of Green's functions when a four momentum p^2 is complex. For example when we obtain the quantity $\ln(p^2/M^2)$ in a Euclidean calculation we claim that the contribution to the Green's function for real Minkowski momenta is simply the analytic continuation of the logarithm from negative to positive argument. However, it is not always possible to rely on this appealing line of reasoning. There may be subtleties that are escaping notice. We would like, for example, to know what are the Minkowskian analogues of the Euclidean exceptional momenta that we have used in our discussions. In other words we would like to know what the singularities of Green's functions are for real Minkowskian momenta. These will be the analogues of the infrared singularities at Euclidean exceptional momenta. Ruelle[25] has shown that for Minkowskian momenta the Green's functions can only be singular when a nontrivial momentum lies on the mass spectrum of the theory. Further when this is the case a singularity only arises for lightlike momenta. In summary, the Euclidean exceptional momenta, (vanishing partial sums), are replaced by lightlike partial sums in Minkowski space. Even with this result it is difficult to obtain general results for Minkowskian momenta. Nevertheless in this connection there are interesting results by Thun[26] for deep inelastic scattering.

Thun[26] introduces a decomposition for the momenta which it is instructive to consider. A momentum p may be expanded in terms of three other momenta a, b, c. Let

$$p = \mu a + b + \frac{c}{\mu} \tag{5.159}$$

and also let

$$\begin{aligned} a^2 = c^2 &= 0 \\ a \cdot b = b \cdot c &= 0, \end{aligned} \tag{5.160}$$

this gives

$$p^2 = 2a \cdot c + b^2. \tag{5.161}$$

This means that as $\mu \to \infty$ some of the momentum components go to infinity but that p^2 remains fixed. This is just what we require for the large-momentum limit of an on shell particle. An example of a choice for a, b and c are the so-called infinite momentum variables,[27] p_+, p_- and $\underline{p}$ where

$$\begin{aligned} p_+ &= p_0 + p_3 = m\mu, \\ p_- &= p_0 - p_3 = \frac{m}{\mu}, \\ \underline{p} &= (0, p_1, p_2, 0). \end{aligned} \tag{5.162}$$

This amounts to the choice

$$a = \frac{m}{2}(1, 0, 0, +1),$$

$$c = \frac{m}{2}(1, 0, 0, -1), \qquad (5.163)$$

$$b = (0, b_1, b_2, 0).$$

If we use these variables in the dimensional identity for Γ we obtain

$$\Gamma\left(\mu a + b + \frac{c}{\mu}, m^2, \lambda\right) = \mu^{4-n}\Gamma\left(a + \frac{b}{\mu} + \frac{c}{\mu^2}, \frac{m^2}{\mu^2}, \lambda\right) \qquad (5.164)$$

then as $\mu \to \infty$, p^2 remains fixed and the momentum on the RHS tends to the lightlike momentum a. This is how the lightlike momenta turn up in Minkowski space. For more details see reference 27.

References

1. S. L. Adler and R. F. Dashen, 'Current Algebras', Benjamin, New York (1968).
2. S. Mandelstam, *in* 'Lectures on Elementary Particles and Quantum Field Theory', (S. Deser, M. Grisaru and H. Pendleton, eds.), MIT Press (1970).
3. G. Mack, 'Strong Interaction Physics', Lecture Notes in Physics, Vol. 17 (W. Ruhl and A. Vancura, eds.), Springer (1972).
4. A. D. Martin and T. D. Spearman, 'Elementary Particle Theory', North-Holland (1970); R. J. Eden, P. V. Landshoff, D. I. Olive and J. C. Polkinghorne, 'The Analytic S-Matrix', Cambridge University Press (1966).
5. J. D. Bjorken, *Phys. Rev.*, **179**, 1547 (1969).
6. K. Wilson, *Phys. Rev.*, **179**, 1499 (1969).
7. R. Brandt and G. Preparata, *Nucl. Phys.*, **B27**, 541 (1971); Y. Frishman, *Phys. Rev. Lett.*, **25**, 966 (1970).
8. H. Fritzsch and M. Gell-Mann, 'International Conference on Duality and Symmetry in Hadron Physics', Weizmann Science Press (1971).
9. C. G. Callan, *Phys. Rev.*, D2, 1541 (1970); K. Symanzik, *Comm. Math. Phys.*, **18**, 227 (1970); G. 't Hooft, *Nucl. Phys.*, **B61**, 455 (1973); S. Weinberg, *Phys. Rev.*, D8, 3497 (1973).
10. D. J. Gross and F. Wilczek, *Phys. Rev. Lett.*, **30**, 1343 (1973); H. D. Politzer, *Phys. Rev. Lett.*, **30**, 1346 (1973); G. 't Hooft, *Nucl. Phys.*, **B61**, 455 (1973).

11. M. Gell-Mann and F. E. Low, *Phys. Rev.*, **95**, 1300 (1954); N. N. Bogoliubov and D. V. Shirkov, 'Introduction to the Theory of Quantised Fields', Interscience, New York (1959).
12. S. Weinberg, *Phys. Rev.*, **118**, 838 (1960).
13. 'Lectures on Elementary Particles and Quantum Field Theory' (S. Deser, M. Grisaru and H. Pendleton, eds.), M.I.T. Press (1970); W. Zimmermann, *Annals of Physics*, **77**, 536, 570 (1973).
14. K. Symanzik, *Comm. Math. Phys.*, **23**, 49 (1971).
15. N. Christ, B. Hasslacher, A. H. Mueller, *Phys. Rev.*, D**6**, 3543 (1972).
16. A. L. Mason, *Phys. Rev.*, D**7**, 3799 (1973).
17. C. Nash, *Nuovo Cimento*, **25A**, 173 (1975).
18. K. Wilson, *Phys. Rev.*, D**3**, 1818 (1971).
19. V. V. Belokurov, D. I. Kazakov, D. V. Shirkov, A. A. Slavnov and A. A. Vladimirov, *Phys. Lett.*, **47B**, 359 (1973).
20. S. Coleman and D. Gross, *Phys. Rev. Lett.*, **31**, 851 (1973).
21. D. Gross and F. Wilczek, *Phys. Rev.*, D**8**, 3633 (1973).
22. K. Symanzik, *Comm. Math. Phys.*, **18**, 227 (1970).
23. G 't Hooft, *Nucl. Phys.*, B**61**, 455 (1973); S. Weinberg, *Phys. Rev.*, D**8**, 3497 (1973).
24. J. C. Collins and A. J. McFarlane, *Phys. Rev.*, D**10**, 1201 (1974).
25. D. Ruelle, *Nuovo Cimento*, **19**, 356 (1961).
26. H. J. Thun, *Nuovo Cimento*, **26A**, 329 (1975).
27. J. Kogut and D. Soper, *Phys. Rev.*, D**1**, 2901 (1970).
28. S. Humble, 'Introduction to Particle Production in Hadron Physics', Academic Press (1974).

Appendix

We work in a system of units in which $\hbar = c = 1$. The metric tensor $g_{\mu\nu}$ is given by (indices run from 0 to 3)

$$g_{\mu\nu} = \begin{pmatrix} 1 & 0 & 0 & 0 \\ 0 & -1 & 0 & 0 \\ 0 & 0 & -1 & 0 \\ 0 & 0 & 0 & -1 \end{pmatrix}. \quad (A.1)$$

The Dirac equation is written in the form

$$(i\gamma^\mu \partial_\mu - m)\psi(x) = 0. \quad (A.2)$$

The Dirac γ matrices γ_μ have the properties

$$\{\gamma_\mu, \gamma_\nu\} = \gamma_\mu \gamma_\nu + \gamma_\nu \gamma_\mu = 2g_{\mu\nu} I \quad (A.3)$$

where I is a 4×4 unit matrix. An explicit form for the Dirac matrices is given by

$$\gamma^0 = \begin{pmatrix} I & 0 \\ 0 & -I \end{pmatrix}, \quad \gamma^i = \begin{pmatrix} 0 & \sigma^i \\ -\sigma^i & 0 \end{pmatrix} \quad (A.4)$$

where I is a 2×2 unit matrix and σ^i are the Pauli matrices defined by

$$\sigma^1 = \begin{pmatrix} 0 & 1 \\ 1 & 0 \end{pmatrix}, \quad \sigma^2 = \begin{pmatrix} 0 & -i \\ i & 0 \end{pmatrix}, \quad \sigma^3 = \begin{pmatrix} 1 & 0 \\ 0 & -1 \end{pmatrix}. \quad (A.5)$$

APPENDIX

An abbreviation that we shall use is $\not{K}$ which stands for $\gamma^\mu k_\mu$. The properties of the γ^μ under Hermitian conjugation are

$$(\gamma^0)^+ = \gamma^0, \qquad (\gamma^i)^+ = -\gamma^i. \tag{A.6}$$

It is also convenient to note that

$$(\gamma^0)^2 = 1, \qquad (\gamma^i)^2 = -1. \tag{A.7}$$

The matrices γ^5 and $\sigma_{\mu\nu}$ defined below are frequently used in calculations

$$\gamma^5 = i\gamma^0\gamma^1\gamma^2\gamma^3$$
$$\sigma_{\mu\nu} = \frac{i}{2}(\gamma_\mu\gamma_\nu - \gamma_\nu\gamma_\mu) = \frac{i}{2}[\gamma_\mu, \gamma_\nu], \qquad \mu, \nu = 0, 1, 2, 3. \tag{A.8}$$

γ^5 has the properties

$$(\gamma^5)^+ = \gamma^5, \qquad (\gamma^5)^2 = 1, \qquad \gamma^5\gamma^\mu = -\gamma^\mu\gamma^5. \tag{A.9}$$

If we write solutions to the Dirac equation as

$$\psi(x) = u(p)e^{-ipx} + v(p)e^{ipx}, \qquad p_0 > 0 \tag{A.10}$$

then in momentum space, the spinors u and v satisfy the equations

$$(\not{p} - m)u = 0, \qquad (\not{p} + m)v = 0. \tag{A.11}$$

The spinor $\bar{u}$ is defined by

$$\bar{u}(p) = u^+\gamma^0. \tag{A.12}$$

If we use the form A.4 for the γ^μ we find that there are two linearly independent u's and two linearly independent v's given by

$$u_+ = \begin{pmatrix} 1 \\ 0 \\ 0 \\ 0 \end{pmatrix} \qquad v_+ = \begin{pmatrix} 0 \\ 0 \\ 1 \\ 0 \end{pmatrix}$$
$$u_- = \begin{pmatrix} 0 \\ 1 \\ 0 \\ 0 \end{pmatrix} \qquad v_- = \begin{pmatrix} 0 \\ 0 \\ 0 \\ 1 \end{pmatrix} \tag{A.13}$$

The label $\mp$ which distinguishes $u_\mp$ and $v_\mp$ is written s and is the spin of the particle. The polarization vector s_μ has the form, in the rest frame of the particle,

$$s_\mu = (0, \mathbf{s}) \tag{A.14}$$

where $\mathbf{s}$ is a unit vector pointing in the direction of the spin of the particle. In terms of s_μ the spinors $u_\mp$, $v_\mp$ satisfy

$$(1 \mp \gamma_5 \slashed{s})u_\pm = 0, \qquad (1 \mp \gamma_5 \slashed{s})v_\pm = 0. \tag{A.15}$$

In terms of u and $\bar{u}$ the operator which projects out positive energy and spin+ is given by

$$u_\alpha \bar{u}_\beta = \left\{ \frac{(\slashed{p}+m)}{2m} \frac{(1+\gamma_5 \slashed{s})}{2} \right\}_{\alpha\beta}. \tag{A.16}$$

The projection operator which projects out only positive energy is obtained by averaging A.14 over the spins$\mp$ and is given by

$$\sum_{s=\mp} u_\alpha \bar{u}_\beta = \left\{ \frac{(\slashed{p}+m)}{2m} \right\}_{\alpha\beta}. \tag{A.17}$$

In coordinate space the wave function of the photon is defined by

$$A_\mu(x) = \int \frac{d^3k}{(2\pi)^3 \sqrt{2k_0}} \sum_{\lambda=0}^{3} \varepsilon_\mu(\mathbf{k}, \lambda) \{a(\mathbf{k}, \lambda) e^{-ik \cdot x} + a^+(\mathbf{k}, \lambda) e^{ik \cdot x}\}$$

where the $\varepsilon_\mu(\mathbf{k}, \lambda)$, $\lambda = 0, \ldots 3$, are the four polarization vectors of the photon. It is usual to choose a coordinate system such that

$$k_\mu = (|\dot{\mathbf{k}}|, |\mathbf{k}|, 0, 0). \tag{A.18}$$

Then the physical polarizations of the photon are $\varepsilon_\mu(\mathbf{k}, 1)$ and $\varepsilon_\mu(\mathbf{k}, 2)$ and they satisfy

$$k^\mu \varepsilon_\mu(\mathbf{k}, 1) = k^\mu \varepsilon_\mu(\mathbf{k}, 2) = 0 \tag{A.19}$$

while the unphysical polarizations satisfy

$$k^\mu \varepsilon_\mu(\mathbf{k}, 0) = -ik^\mu \varepsilon_\mu(\mathbf{k}, 3) = |\mathbf{k}|. \tag{A.20}$$

An explicit form for the $\varepsilon_\mu(\mathbf{k}, \lambda)$ is then

$$\begin{aligned}
\varepsilon_\mu(\mathbf{k}, 0) &= (i, 0, 0, 0), \\
\varepsilon_\mu(\mathbf{k}, 1) &= (0, 0, 1, 0), \\
\varepsilon_\mu(\mathbf{k}, 2) &= (0, 0, 0, 1), \\
\varepsilon_\mu(\mathbf{k}, 3) &= (0, 1, 0, 0).
\end{aligned} \tag{A.21}$$

APPENDIX

Then the $\varepsilon_\mu(\mathbf{k}, \lambda)$ also satisfy the equation

$$\sum_{\lambda=0}^{3} \varepsilon_\mu(\mathbf{k}, \lambda)\varepsilon_\nu(\mathbf{k}, \lambda) = g_{\mu\nu}. \quad (A.22)$$

The Feynman rules for $\lambda\phi^4/4!$, QED and Yang-Mills theory are to be found in Chapter 2, p. 63, Chapter 3, p. 92 and Chapter 5, pp. 196–197, respectively.

Some useful formulae in evaluating integrals are

volume element in n-dimensional space

$$d^n k = dk\, k^{n-1} \sin^{n-2}\theta_{n-1} \sin^{n-3}\theta_{n-2} \ldots d\theta_{n-1} \ldots d\theta, \quad (A.23)$$

$k = \sqrt{k^2}$, $0 \le \theta_1 \le 2\pi$ and all other angles range from 0 to π.

$$\int \frac{d^n k}{[k^2 + b^2]^p} = i\pi^{n/2} \frac{\Gamma(p - n/2)}{\Gamma(p)} \frac{1}{(b^2)^{p-n/2}} \quad (A.24)$$

$$\int \frac{d^n k\, k_\mu k_\nu}{[k^2 + b^2]^p} = \frac{i\pi^{n/2}}{\Gamma(p)} \frac{1}{(b^2)^{p-(n/2)-1}} \cdot \frac{\Gamma(p - (n/2) - 1)}{2} g_{\mu\nu}. \quad (A.25)$$

For further information and detailed proofs of the properties of γ matrices and relativistic kinematics the reader may consult the following books

J. D. Bjorken and S. D. Drell, 'Relativistic Quantum Mechanics', McGraw-Hill, 1964.

S. Schweber, 'An Introduction to Relativistic Quantum Field Theory', Harper and Row, New York, 1961.

Bibliography

F. A. Berezin, "Method of Second Quantization", Academic Press, New York and London (1966).
  This is a useful reference text which discusses functional methods for Bose and Fermi systems.
S. A. Albeverio and R. J. Høegh-Krohn, "Mathematical Theory of Feynman Integrals", Springer Lecture notes in Physics No. 523, Springer-Verlag, Berlin (1976).
  This book contains material of interest to those who wish to place the Feynman integral on a rigorous mathematical basis.
G. Iverson, A. Perlmutter and S. Minty, (editors), "Fundamental Interactions in Physics and Astrophysics", Plenum Press N.Y. (1973).
  The article by B. Simon in this book contains a lucid summary of what was known about the divergence of the perturbation series in various theories in 1973. Since then there has been considerable further work, e.g. J. P. Eckmann, J. Magnen and R. Seneor, *Comm. Math. Phys.* **39**, 251 (1975); J. S. Feldman and K. Osterwalder, *Ann. Phys.* **97**, 80 (1976). Additional papers continue to appear.
C. G. Bollini and J. H. Giambiagi, *Physics Lett.* **40B**, 566 (1972). J. C. Collins, *Nucl. Phys.* **B92**, 477 (1975).
  These two articles are further references which can be profitably read after digesting chapter two and its references.
J. C. Taylor, "Gauge Theories of the Weak Interactions", Cambridge University Press, Cambridge (1976).
  This book is an excellent introduction to the properties and applications of non-Abelian gauge theories in physics.
B. Lautrup, *Nucl. Phys.* **B105**, 23 (1976).
  This is a very good article to read after mastering the material in Chapters

3 and 5. It does many important computations for QED using the dimensional method, and considerably extends the results obtained in Chapter 3.

R. F. Streater, Outline of Axiomatic Relativistic Quantum Field Theory; *Rep. Prog. Phys.* **38**, 771 (1975).

This is a review covering the work done in axiomatic field theory from 1954–74. It is very thorough and contains a good set of references. In particular it describes the recent work on Euclidean field theory.

Index

A

Action principle, 2
Analytic continuation in D, 70–77, 97–98
  lack of uniqueness and Ward identity, 97–98
  for multiloop diagrams, 75–77
  for single loop diagrams, 72–75
Analytic continuation in k^2, 66
Anomalous dimension, 193
Asymptotic behaviour of Green's functions, 165–166, 175–188, 190–195, 202, 207–212
  and the function β, 190–195
  corrections to, 207–210
  for exceptional momenta, 187–188
  for non-exceptional momenta, 177–181
  for Yang–Mills theories, 202
Asymptotic expansion, 62
Asymptotic freedom, 194–195, 202

B

Brownian motion, 25–26

C

Callan–Symanzik equations 166–204
  for $\lambda\phi^4/4!$ theory, 168
  for QED, 203
  for Yang–Mills theories, 198–202
  for zero mass theories, 198–199, 203–204
Coefficient functions, 170–175, 198–202
  for Callan–Symanzik equations, 170–175
  for 't Hooft–Weinberg equations, 198–202
Coherent states, 150–153
Conditional convergence, 66
Connected diagrams, see also Green's functions, connected, 38
Counter term, 67, 69, 79–85, 90, 107, 113, 117

D

Degree of divergence, 86–87, 93–95
  in $\lambda\phi^4/4!$ theory, 86–87
  in QED, 93–95
Diffusion equation, 25, 30
  with imaginary coefficient 30
Dimension of Feynman graph, 86, 93, 172
  in $\lambda\phi^4/4!$ theory, 86, 172
  in QED, 93
D-dimensional momenta, 73
Dimension of operators, 189
Dimensional regularization, 61–121
  of $\lambda\phi^4/4!$ theory, 70–85, 89–90
  of QED, 100–118
Dirac matrices, 101–102, 108–109
  traces of, 101–102

INDEX

E

Effective coupling constant, 179–180, 202
  for Yang–Mills theories, 202
Effective mass, 206–209
Euclidean momenta, 66
Exceptional momenta, see also asymptotic behaviour, 175–177, 182–188, 208–209, 210–212
  and inclusive processes, 176–177
  and infrared divergences, 182–185, 208
  Minkowskian, 210–212

F

Feynman gauge, 135
Feynman rules, 3, 10, 63, 92, 140–141, 196–197.
  for $\lambda\phi^4/4!$ theory, 3, 63
  for QED, 10, 92
  for QED with ghosts, 140–141
  for Yang–Mills theories, 196–197
Fixed points, 166, 190–194
Fredholm determinant, 27–29, 48, 128
Functional derivative, 15–24
Functional Fourier transform, 55
Functional integrals, 24–49
Functional power series, 18–19, 22–23
Functional Taylor series, 23–24

G

Gauge invariance, see also Ward identity, 91, 94, 106–107, 125–135
Gauge fixing term, 92, 125–135, 139–141
Gauge transformation, 127–135
  for Yang–Mills theories, 132
Gell-Mann–Low equation, 165–166
Generating functions, 50–55, 136–138
  for amputated one particle irreducible Green's functions, 51–55, 136–138
  for connected Green's functions, 50–51, 136–137
Ghost, 129–130, 140–141, 196–197
Grassmann algebra, 19–23, 45–47
  change of variables in, 46
  differentiation in, 21–23
  of finite dimension, 20
  of infinite dimension, 21–23
  integration on, 45–47
Green's functions, 35–44, 48–49, 50–55, 167
  amputated, 38
  connected, 38
  for free particle, 35–37
  for interacting fields, 37
  one particle irreducible, 51–55, 167
  for QED, 48–49

H

Hard photons, 154–155

I

Infinite momentum variables, 211–212
Infrared divergence cancellation in QED, 148–150, 153–161
Infrared divergence and renormalization of QED, 111–113, 115–117
Infrared stability, 190–194
Integration by parts, 73–75, 76–77
Intermediate renormalization, 112–113

L

Lagrangian, 63, 91, 195
  for $\lambda\phi^4/4!$ theory, 63
  for QED, 91
  for Yang–Mills theories, 195
Landau gauge, 135
Legendre transform, 52, 137
Linear divergence, 77
Logarithmic divergence, 66, 69

M

Mandelstam invariants, 67
Mass shift see renormalized mass

N

Negative probability, 143–144, 148
Non-renormalizable theory, 121–124
Normal product, 43–44, 49
  for Fermions, 49

O

Overlapping divergence, 78–85

P

Path integral, 29–30

INDEX

Polarization vectors for photons, 144–145, 147
Pole, 71, 80, 83, 85
 double, 83, 85
 multiple, 85
 residue of, 80–83
 simple, 71
Propagator, 3–6, 11–12, 39
 for electron, 11–12
 inverse, 39
 in $\lambda\phi^4/4!$ theory, 3–6

Q
Quadratic divergence, 77–79, 81–85

R
Renormalization constants, 88, 117
 for $\lambda\phi^4/4!$ theory, 88
 for QED, 117
Renormalization group methods, see also names of specific equations, 164–212
Renormalization in general, 86–90, 121–124
 criteria for, 121–124
Renormalization of $\lambda\phi^4/4!$ theory, 3–9, 67–90
Renormalization of QED, 9–15, 67–118
Renormalized coupling constant, 8, 14, 88, 118
 in $\lambda\phi^4/4!$ theory, 8, 88
 in QED, 14, 118
Renormalized field, 6, 88, 118
 for electron, 118
 in $\lambda\phi^4/4!$ theory, 6, 88
 for photon, 118
Renormalized mass, 4–5, 11–12, 85, 88, 117–118
 for electron, 11–12, 117–118
 in $\lambda\phi^4/4!$ theory, 4–5, 85, 88
Renormalized vertex see renormalized coupling constant

S
Schrödinger equation, 30

Schwinger–Dyson equations, 55–60
Soft photons, 150, 154–161
 cross-section for, 161
Source, 31–32
Spectral representation, 142–143, 188
Spence function, 116
Super-renormalizable theory, 123–124
Symmetry factor, 3, 63–65
Symmetry point, 7, 171, 173

T
't Hooft–Weinberg equations, 204–209
Threshold branch cut, 70
Triangle anomaly, 118–121

U
Ultraviolet cut-off, 66–69
Ultraviolet stability, 190–194
Unitarity, 70, 89–90, 141–148
 and insertion of counter terms, 89–90
 and threshold branch cut, 70
 and Ward identity, 145–148

V
Vacuum graphs, 41–44
Vacuum, polarization, 104–108
Variational equation, 2, 9
 for electron, 9
 for photon, 9
 for scalar field, 2

W
Ward identity, 95–100, 115, 120, 135–139, 141, 148
Wave function renormalization, 5–6, 10–13, 85, 88, 117–118
 for electron, 11–13, 117–118
 in $\lambda\phi^4/4!$ theory, 5–6, 85, 88
 for photon, 10–11, 117–118
Weinberg's theorem, 87
Wick products, 42
Wiener measure, 26–30
Wilson expansion, 185, 188–190
 and exceptional momenta, 185

A CATALOG OF SELECTED
DOVER BOOKS
IN ALL FIELDS OF INTEREST

A CATALOG OF SELECTED DOVER BOOKS IN ALL FIELDS OF INTEREST

100 BEST-LOVED POEMS, Edited by Philip Smith. "The Passionate Shepherd to His Love," "Shall I compare thee to a summer's day?" "Death, be not proud," "The Raven," "The Road Not Taken," plus works by Blake, Wordsworth, Byron, Shelley, Keats, many others. 96pp. 5 3/16 x 8 1/4. 0-486-28553-7

100 SMALL HOUSES OF THE THIRTIES, Brown-Blodgett Company. Exterior photographs and floor plans for 100 charming structures. Illustrations of models accompanied by descriptions of interiors, color schemes, closet space, and other amenities. 200 illustrations. 112pp. 8 3/8 x 11. 0-486-44131-8

1000 TURN-OF-THE-CENTURY HOUSES: With Illustrations and Floor Plans, Herbert C. Chivers. Reproduced from a rare edition, this showcase of homes ranges from cottages and bungalows to sprawling mansions. Each house is meticulously illustrated and accompanied by complete floor plans. 256pp. 9 3/8 x 12 1/4.
0-486-45596-3

101 GREAT AMERICAN POEMS, Edited by The American Poetry & Literacy Project. Rich treasury of verse from the 19th and 20th centuries includes works by Edgar Allan Poe, Robert Frost, Walt Whitman, Langston Hughes, Emily Dickinson, T. S. Eliot, other notables. 96pp. 5 3/16 x 8 1/4. 0-486-40158-8

101 GREAT SAMURAI PRINTS, Utagawa Kuniyoshi. Kuniyoshi was a master of the warrior woodblock print — and these 18th-century illustrations represent the pinnacle of his craft. Full-color portraits of renowned Japanese samurais pulse with movement, passion, and remarkably fine detail. 112pp. 8 3/8 x 11. 0-486-46523-3

ABC OF BALLET, Janet Grosser. Clearly worded, abundantly illustrated little guide defines basic ballet-related terms: arabesque, battement, pas de chat, relevé, sissonne, many others. Pronunciation guide included. Excellent primer. 48pp. 4 3/16 x 5 3/4.
0-486-40871-X

ACCESSORIES OF DRESS: An Illustrated Encyclopedia, Katherine Lester and Bess Viola Oerke. Illustrations of hats, veils, wigs, cravats, shawls, shoes, gloves, and other accessories enhance an engaging commentary that reveals the humor and charm of the many-sided story of accessorized apparel. 644 figures and 59 plates. 608pp. 6 1/8 x 9 1/4.
0-486-43378-1

ADVENTURES OF HUCKLEBERRY FINN, Mark Twain. Join Huck and Jim as their boyhood adventures along the Mississippi River lead them into a world of excitement, danger, and self-discovery. Humorous narrative, lyrical descriptions of the Mississippi valley, and memorable characters. 224pp. 5 3/16 x 8 1/4. 0-486-28061-6

ALICE STARMORE'S BOOK OF FAIR ISLE KNITTING, Alice Starmore. A noted designer from the region of Scotland's Fair Isle explores the history and techniques of this distinctive, stranded-color knitting style and provides copious illustrated instructions for 14 original knitwear designs. 208pp. 8 3/8 x 10 7/8. 0-486-47218-3

Browse over 9,000 books at www.doverpublications.com

CATALOG OF DOVER BOOKS

ALICE'S ADVENTURES IN WONDERLAND, Lewis Carroll. Beloved classic about a little girl lost in a topsy-turvy land and her encounters with the White Rabbit, March Hare, Mad Hatter, Cheshire Cat, and other delightfully improbable characters. 42 illustrations by Sir John Tenniel. 96pp. 5 3/16 x 8 1/4. 0-486-27543-4

AMERICA'S LIGHTHOUSES: An Illustrated History, Francis Ross Holland. Profusely illustrated fact-filled survey of American lighthouses since 1716. Over 200 stations — East, Gulf, and West coasts, Great Lakes, Hawaii, Alaska, Puerto Rico, the Virgin Islands, and the Mississippi and St. Lawrence Rivers. 240pp. 8 x 10 3/4. 0-486-25576-X

AN ENCYCLOPEDIA OF THE VIOLIN, Alberto Bachmann. Translated by Frederick H. Martens. Introduction by Eugene Ysaye. First published in 1925, this renowned reference remains unsurpassed as a source of essential information, from construction and evolution to repertoire and technique. Includes a glossary and 73 illustrations. 496pp. 6 1/8 x 9 1/4. 0-486-46618-3

ANIMALS: 1,419 Copyright-Free Illustrations of Mammals, Birds, Fish, Insects, etc., Selected by Jim Harter. Selected for its visual impact and ease of use, this outstanding collection of wood engravings presents over 1,000 species of animals in extremely lifelike poses. Includes mammals, birds, reptiles, amphibians, fish, insects, and other invertebrates. 284pp. 9 x 12. 0-486-23766-4

THE ANNALS, Tacitus. Translated by Alfred John Church and William Jackson Brodribb. This vital chronicle of Imperial Rome, written by the era's great historian, spans A.D. 14-68 and paints incisive psychological portraits of major figures, from Tiberius to Nero. 416pp. 5 3/16 x 8 1/4. 0-486-45236-0

ANTIGONE, Sophocles. Filled with passionate speeches and sensitive probing of moral and philosophical issues, this powerful and often-performed Greek drama reveals the grim fate that befalls the children of Oedipus. Footnotes. 64pp. 5 3/16 x 8 1/4. 0-486-27804-2

ART DECO DECORATIVE PATTERNS IN FULL COLOR, Christian Stoll. Reprinted from a rare 1910 portfolio, 160 sensuous and exotic images depict a breathtaking array of florals, geometrics, and abstracts — all elegant in their stark simplicity. 64pp. 8 3/8 x 11. 0-486-44862-2

THE ARTHUR RACKHAM TREASURY: 86 Full-Color Illustrations, Arthur Rackham. Selected and Edited by Jeff A. Menges. A stunning treasury of 86 full-page plates span the famed English artist's career, from *Rip Van Winkle* (1905) to masterworks such as *Undine, A Midsummer Night's Dream,* and *Wind in the Willows* (1939). 96pp. 8 3/8 x 11. 0-486-44685-9

THE AUTHENTIC GILBERT & SULLIVAN SONGBOOK, W. S. Gilbert and A. S. Sullivan. The most comprehensive collection available, this songbook includes selections from every one of Gilbert and Sullivan's light operas. Ninety-two numbers are presented uncut and unedited, and in their original keys. 410pp. 9 x 12. 0-486-23482-7

THE AWAKENING, Kate Chopin. First published in 1899, this controversial novel of a New Orleans wife's search for love outside a stifling marriage shocked readers. Today, it remains a first-rate narrative with superb characterization. New introductory Note. 128pp. 5 3/16 x 8 1/4. 0-486-27786-0

BASIC DRAWING, Louis Priscilla. Beginning with perspective, this commonsense manual progresses to the figure in movement, light and shade, anatomy, drapery, composition, trees and landscape, and outdoor sketching. Black-and-white illustrations throughout. 128pp. 8 3/8 x 11. 0-486-45815-6

CATALOG OF DOVER BOOKS

THE BATTLES THAT CHANGED HISTORY, Fletcher Pratt. Historian profiles 16 crucial conflicts, ancient to modern, that changed the course of Western civilization. Gripping accounts of battles led by Alexander the Great, Joan of Arc, Ulysses S. Grant, other commanders. 27 maps. 352pp. 5⅜ x 8½.   0-486-41129-X

BEETHOVEN'S LETTERS, Ludwig van Beethoven. Edited by Dr. A. C. Kalischer. Features 457 letters to fellow musicians, friends, greats, patrons, and literary men. Reveals musical thoughts, quirks of personality, insights, and daily events. Includes 15 plates. 410pp. 5⅜ x 8½.   0-486-22769-3

BERNICE BOBS HER HAIR AND OTHER STORIES, F. Scott Fitzgerald. This brilliant anthology includes 6 of Fitzgerald's most popular stories: "The Diamond as Big as the Ritz," the title tale, "The Offshore Pirate," "The Ice Palace," "The Jelly Bean," and "May Day." 176pp. 5⅜ x 8½.   0-486-47049-0

BESLER'S BOOK OF FLOWERS AND PLANTS: 73 Full-Color Plates from Hortus Eystettensis, 1613, Basilius Besler. Here is a selection of magnificent plates from the *Hortus Eystettensis,* which vividly illustrated and identified the plants, flowers, and trees that thrived in the legendary German garden at Eichstätt. 80pp. 8⅜ x 11.
0-486-46005-3

THE BOOK OF KELLS, Edited by Blanche Cirker. Painstakingly reproduced from a rare facsimile edition, this volume contains full-page decorations, portraits, illustrations, plus a sampling of textual leaves with exquisite calligraphy and ornamentation. 32 full-color illustrations. 32pp. 9⅜ x 12¼.   0-486-24345-1

THE BOOK OF THE CROSSBOW: With an Additional Section on Catapults and Other Siege Engines, Ralph Payne-Gallwey. Fascinating study traces history and use of crossbow as military and sporting weapon, from Middle Ages to modern times. Also covers related weapons: balistas, catapults, Turkish bows, more. Over 240 illustrations. 400pp. 7¼ x 10⅛.   0-486-28720-3

THE BUNGALOW BOOK: Floor Plans and Photos of 112 Houses, 1910, Henry L. Wilson. Here are 112 of the most popular and economic blueprints of the early 20th century — plus an illustration or photograph of each completed house. A wonderful time capsule that still offers a wealth of valuable insights. 160pp. 8⅜ x 11.
0-486-45104-6

THE CALL OF THE WILD, Jack London. A classic novel of adventure, drawn from London's own experiences as a Klondike adventurer, relating the story of a heroic dog caught in the brutal life of the Alaska Gold Rush. Note. 64pp. 5³⁄₁₆ x 8¼.
0-486-26472-6

CANDIDE, Voltaire. Edited by Francois-Marie Arouet. One of the world's great satires since its first publication in 1759. Witty, caustic skewering of romance, science, philosophy, religion, government — nearly all human ideals and institutions. 112pp. 5³⁄₁₆ x 8¼.   0-486-26689-3

CELEBRATED IN THEIR TIME: Photographic Portraits from the George Grantham Bain Collection, Edited by Amy Pastan. With an Introduction by Michael Carlebach. Remarkable portrait gallery features 112 rare images of Albert Einstein, Charlie Chaplin, the Wright Brothers, Henry Ford, and other luminaries from the worlds of politics, art, entertainment, and industry. 128pp. 8⅜ x 11.   0-486-46754-6

CHARIOTS FOR APOLLO: The NASA History of Manned Lunar Spacecraft to 1969, Courtney G. Brooks, James M. Grimwood, and Loyd S. Swenson, Jr. This illustrated history by a trio of experts is the definitive reference on the Apollo spacecraft and lunar modules. It traces the vehicles' design, development, and operation in space. More than 100 photographs and illustrations. 576pp. 6¾ x 9¼. 0-486-46756-2

Browse over 9,000 books at www.doverpublications.com

CATALOG OF DOVER BOOKS

A CHRISTMAS CAROL, Charles Dickens. This engrossing tale relates Ebenezer Scrooge's ghostly journeys through Christmases past, present, and future and his ultimate transformation from a harsh and grasping old miser to a charitable and compassionate human being. 80pp. 5 3/16 x 8 1/4. 0-486-26865-9

COMMON SENSE, Thomas Paine. First published in January of 1776, this highly influential landmark document clearly and persuasively argued for American separation from Great Britain and paved the way for the Declaration of Independence. 64pp. 5 3/16 x 8 1/4. 0-486-29602-4

THE COMPLETE SHORT STORIES OF OSCAR WILDE, Oscar Wilde. Complete texts of "The Happy Prince and Other Tales," "A House of Pomegranates," "Lord Arthur Savile's Crime and Other Stories," "Poems in Prose," and "The Portrait of Mr. W. H." 208pp. 5 3/16 x 8 1/4. 0-486-45216-6

COMPLETE SONNETS, William Shakespeare. Over 150 exquisite poems deal with love, friendship, the tyranny of time, beauty's evanescence, death, and other themes in language of remarkable power, precision, and beauty. Glossary of archaic terms. 80pp. 5 3/16 x 8 1/4. 0-486-26686-9

THE COUNT OF MONTE CRISTO: Abridged Edition, Alexandre Dumas. Falsely accused of treason, Edmond Dantès is imprisoned in the bleak Chateau d'If. After a hair-raising escape, he launches an elaborate plot to extract a bitter revenge against those who betrayed him. 448pp. 5 3/16 x 8 1/4. 0-486-45643-9

CRAFTSMAN BUNGALOWS: Designs from the Pacific Northwest, Yoho & Merritt. This reprint of a rare catalog, showcasing the charming simplicity and cozy style of Craftsman bungalows, is filled with photos of completed homes, plus floor plans and estimated costs. An indispensable resource for architects, historians, and illustrators. 112pp. 10 x 7. 0-486-46875-5

CRAFTSMAN BUNGALOWS: 59 Homes from "The Craftsman," Edited by Gustav Stickley. Best and most attractive designs from Arts and Crafts Movement publication — 1903–1916 — includes sketches, photographs of homes, floor plans, descriptive text. 128pp. 8 1/4 x 11. 0-486-25829-7

CRIME AND PUNISHMENT, Fyodor Dostoyevsky. Translated by Constance Garnett. Supreme masterpiece tells the story of Raskolnikov, a student tormented by his own thoughts after he murders an old woman. Overwhelmed by guilt and terror, he confesses and goes to prison. 480pp. 5 3/16 x 8 1/4. 0-486-41587-2

THE DECLARATION OF INDEPENDENCE AND OTHER GREAT DOCUMENTS OF AMERICAN HISTORY: 1775-1865, Edited by John Grafton. Thirteen compelling and influential documents: Henry's "Give Me Liberty or Give Me Death," Declaration of Independence, The Constitution, Washington's First Inaugural Address, The Monroe Doctrine, The Emancipation Proclamation, Gettysburg Address, more. 64pp. 5 3/16 x 8 1/4. 0-486-41124-0

THE DESERT AND THE SOWN: Travels in Palestine and Syria, Gertrude Bell. "The female Lawrence of Arabia," Gertrude Bell wrote captivating, perceptive accounts of her travels in the Middle East. This intriguing narrative, accompanied by 160 photos, traces her 1905 sojourn in Lebanon, Syria, and Palestine. 368pp. 5 3/8 x 8 1/2. 0-486-46876-3

A DOLL'S HOUSE, Henrik Ibsen. Ibsen's best-known play displays his genius for realistic prose drama. An expression of women's rights, the play climaxes when the central character, Nora, rejects a smothering marriage and life in "a doll's house." 80pp. 5 3/16 x 8 1/4. 0-486-27062-9

Browse over 9,000 books at www.doverpublications.com

CATALOG OF DOVER BOOKS

DOOMED SHIPS: Great Ocean Liner Disasters, William H. Miller, Jr. Nearly 200 photographs, many from private collections, highlight tales of some of the vessels whose pleasure cruises ended in catastrophe: the *Morro Castle, Normandie, Andrea Doria, Europa,* and many others. 128pp. 8⅞ x 11¾. 0-486-45366-9

THE DORÉ BIBLE ILLUSTRATIONS, Gustave Doré. Detailed plates from the Bible: the Creation scenes, Adam and Eve, horrifying visions of the Flood, the battle sequences with their monumental crowds, depictions of the life of Jesus, 241 plates in all. 241pp. 9 x 12. 0-486-23004-X

DRAWING DRAPERY FROM HEAD TO TOE, Cliff Young. Expert guidance on how to draw shirts, pants, skirts, gloves, hats, and coats on the human figure, including folds in relation to the body, pull and crush, action folds, creases, more. Over 200 drawings. 48pp. 8¼ x 11. 0-486-45591-2

DUBLINERS, James Joyce. A fine and accessible introduction to the work of one of the 20th century's most influential writers, this collection features 15 tales, including a masterpiece of the short-story genre, "The Dead." 160pp. 5 3/16 x 8¼.
0-486-26870-5

EASY-TO-MAKE POP-UPS, Joan Irvine. Illustrated by Barbara Reid. Dozens of wonderful ideas for three-dimensional paper fun — from holiday greeting cards with moving parts to a pop-up menagerie. Easy-to-follow, illustrated instructions for more than 30 projects. 299 black-and-white illustrations. 96pp. 8⅜ x 11.
0-486-44622-0

EASY-TO-MAKE STORYBOOK DOLLS: A "Novel" Approach to Cloth Dollmaking, Sherralyn St. Clair. Favorite fictional characters come alive in this unique beginner's dollmaking guide. Includes patterns for Pollyanna, Dorothy from *The Wonderful Wizard of Oz,* Mary of *The Secret Garden,* plus easy-to-follow instructions, 263 black-and-white illustrations, and an 8-page color insert. 112pp. 8¼ x 11. 0-486-47360-0

EINSTEIN'S ESSAYS IN SCIENCE, Albert Einstein. Speeches and essays in accessible, everyday language profile influential physicists such as Niels Bohr and Isaac Newton. They also explore areas of physics to which the author made major contributions. 128pp. 5 x 8. 0-486-47011-3

EL DORADO: Further Adventures of the Scarlet Pimpernel, Baroness Orczy. A popular sequel to *The Scarlet Pimpernel,* this suspenseful story recounts the Pimpernel's attempts to rescue the Dauphin from imprisonment during the French Revolution. An irresistible blend of intrigue, period detail, and vibrant characterizations. 352pp. 5 3/16 x 8¼. 0-486-44026-5

ELEGANT SMALL HOMES OF THE TWENTIES: 99 Designs from a Competition, Chicago Tribune. Nearly 100 designs for five- and six-room houses feature New England and Southern colonials, Normandy cottages, stately Italianate dwellings, and other fascinating snapshots of American domestic architecture of the 1920s. 112pp. 9 x 12. 0-486-46910-7

THE ELEMENTS OF STYLE: The Original Edition, William Strunk, Jr. This is the book that generations of writers have relied upon for timeless advice on grammar, diction, syntax, and other essentials. In concise terms, it identifies the principal requirements of proper style and common errors. 64pp. 5⅜ x 8½. 0-486-44798-7

THE ELUSIVE PIMPERNEL, Baroness Orczy. Robespierre's revolutionaries find their wicked schemes thwarted by the heroic Pimpernel — Sir Percival Blakeney. In this thrilling sequel, Chauvelin devises a plot to eliminate the Pimpernel and his wife. 272pp. 5 3/16 x 8¼. 0-486-45464-9

Browse over 9,000 books at www.doverpublications.com

CATALOG OF DOVER BOOKS

AN ENCYCLOPEDIA OF BATTLES: Accounts of Over 1,560 Battles from 1479 B.C. to the Present, David Eggenberger. Essential details of every major battle in recorded history from the first battle of Megiddo in 1479 B.C. to Grenada in 1984. List of battle maps. 99 illustrations. 544pp. 6½ x 9¼. 0-486-24913-1

ENCYCLOPEDIA OF EMBROIDERY STITCHES, INCLUDING CREWEL, Marion Nichols. Precise explanations and instructions, clearly illustrated, on how to work chain, back, cross, knotted, woven stitches, and many more — 178 in all, including Cable Outline, Whipped Satin, and Eyelet Buttonhole. Over 1400 illustrations. 219pp. 8⅜ x 11¼. 0-486-22929-7

ENTER JEEVES: 15 Early Stories, P. G. Wodehouse. Splendid collection contains first 8 stories featuring Bertie Wooster, the deliciously dim aristocrat and Jeeves, his brainy, imperturbable manservant. Also, the complete Reggie Pepper (Bertie's prototype) series. 288pp. 5⅜ x 8½. 0-486-29717-9

ERIC SLOANE'S AMERICA: Paintings in Oil, Michael Wigley. With a Foreword by Mimi Sloane. Eric Sloane's evocative oils of America's landscape and material culture shimmer with immense historical and nostalgic appeal. This original hardcover collection gathers nearly a hundred of his finest paintings, with subjects ranging from New England to the American Southwest. 128pp. 10⅜ x 9.
0-486-46525-X

ETHAN FROME, Edith Wharton. Classic story of wasted lives, set against a bleak New England background. Superbly delineated characters in a hauntingly grim tale of thwarted love. Considered by many to be Wharton's masterpiece. 96pp. 5³⁄₁₆ x 8¼.
0-486-26690-7

THE EVERLASTING MAN, G. K. Chesterton. Chesterton's view of Christianity — as a blend of philosophy and mythology, satisfying intellect and spirit — applies to his brilliant book, which appeals to readers' heads as well as their hearts. 288pp. 5⅜ x 8½.
0-486-46036-3

THE FIELD AND FOREST HANDY BOOK, Daniel Beard. Written by a co-founder of the Boy Scouts, this appealing guide offers illustrated instructions for building kites, birdhouses, boats, igloos, and other fun projects, plus numerous helpful tips for campers. 448pp. 5³⁄₁₆ x 8¼. 0-486-46191-2

FINDING YOUR WAY WITHOUT MAP OR COMPASS, Harold Gatty. Useful, instructive manual shows would-be explorers, hikers, bikers, scouts, sailors, and survivalists how to find their way outdoors by observing animals, weather patterns, shifting sands, and other elements of nature. 288pp. 5⅜ x 8½. 0-486-40613-X

FIRST FRENCH READER: A Beginner's Dual-Language Book, Edited and Translated by Stanley Appelbaum. This anthology introduces 50 legendary writers — Voltaire, Balzac, Baudelaire, Proust, more — through passages from *The Red and the Black*, *Les Misérables*, *Madame Bovary*, and other classics. Original French text plus English translation on facing pages. 240pp. 5⅜ x 8½. 0-486-46178-5

FIRST GERMAN READER: A Beginner's Dual-Language Book, Edited by Harry Steinhauer. Specially chosen for their power to evoke German life and culture, these short, simple readings include poems, stories, essays, and anecdotes by Goethe, Hesse, Heine, Schiller, and others. 224pp. 5⅜ x 8½. 0-486-46179-3

FIRST SPANISH READER: A Beginner's Dual-Language Book, Angel Flores. Delightful stories, other material based on works of Don Juan Manuel, Luis Taboada, Ricardo Palma, other noted writers. Complete faithful English translations on facing pages. Exercises. 176pp. 5⅜ x 8½. 0-486-25810-6

Browse over 9,000 books at www.doverpublications.com